AF589383

Smart Plant™

Aligning AI, Automation, and People

By Lisa Ryan

Smart Plant™
Aligning AI, Automation, and People

Published by ®Grategy Press
Cleveland, Ohio

ISBN: 979-8-9951747-0-7
First Edition, 2026
Manufactured in the United States of America

Permissions

All quotes from the *Manufacturers' Network* podcast are used with permission. Company details have been anonymized where requested.

Oakridge Industries is a fictional composite organization created to illustrate patterns observed across multiple manufacturing facilities. No single detail in the Oakridge narrative belongs to any specific company.

The Smart Plant Framework™ is a trademark of Grategy.

ALSO BY LISA RYAN

Manufacturing Engagement: 98 Proven Strategies to Attract and Retain Your Industry's Top Talent

The Upside of Down Times: Discovering the Power of Gratitude

Bestseller. Featured by Harvey Mackay, a nationally syndicated columnist and the *New York Times* bestselling author of *Swim With the Sharks Without Getting Eaten Alive*

Gear Up for Greatness: How to Transform Workplace Culture with the Six Gears of Grategy

Thank You Very Much: Gratitude Strategies to Create a Workplace Culture That ROCKS!

...and eight other titles on employee engagement, gratitude, and workplace culture.

For a complete list, visit LisaRyanSpeaks.com

DEDICATION

For the operators and supervisors who still hear what the sensors miss, and for the leaders trying to build plants where that listening still matters.

Table of Contents

INTRODUCTION: The Pattern

"Leaders are treating digital transformation like a tech install instead of a culture shift. They're not prioritizing people first."

When Vince Sassano, President of Strategic Performance Company, made that statement on the *Manufacturers' Network* podcast, I stopped him there. Not because what he said was controversial, but because it was the clearest articulation of something I had been hearing for years without being able to name.

Between 2020 and 2025, I conducted more than 200 long-form interviews with manufacturing leaders through the *Manufacturers' Network* podcast. CEOs, plant managers, operations directors, supply chain leaders, and the people who actually run the lines. When transcribed, those conversations totaled nearly 1,000,000 words, roughly the length of 10 typical business books.

I wasn't conducting interviews for a book. I was trying to understand what was happening inside manufacturing, beyond conference stages and marketing decks. What people were worried about. What didn't match the narrative. What kept them up at night.

What emerged was a pattern so consistent it became impossible to ignore.

The Pattern I Found

Manufacturers were celebrating automation wins. Higher throughput. Lower costs. Better quality. Everything looked strong on the dashboards. But underneath the metrics, something else was happening. Gradually. Barely noticed. The workforce that really understood the systems – the people who could sense when something was wrong before the data confirmed it – was disappearing.

Not through layoffs. Through erosion. Retirements. Departures. Disengagement. And most dangerously, through systems that were quietly training people to stop using their judgment because the technology was handling the decisions for them.

What makes this failure so dangerous is not that the numbers are wrong. The numbers are right. That's the problem. On paper, your facility might be healthier than ever. But in reality, it may be more vulnerable than at any point in the last fifty years.

Welcome to the paradox of the automated plant: the better your technology performs, the faster you're losing the human foundation that makes survival possible.

Why I'm Writing This Book

I'm not a plant manager. I've never run a manufacturing facility. I'm not an engineer. But I'm not an outsider either.

I spent thirteen years in industrial sales, including seven in the welding industry. I've been inside steel mills, fabrication shops, automotive plants, food processing facilities, and factories of every size. I've walked production floors from the salt mines 1,500 feet below Lake Erie to aerospace facilities building components for space.

What I bring to this conversation isn't operational authority. It's pattern recognition. When you've been in hundreds of plants, you start to see what people inside one facility can't always see. The same mistakes repeated across dozens of facilities. The same erosions happening while dashboards stay green. The same moment when smooth operations quietly become fragile ones.

Here's what I've learned: Most manufacturers have brilliant technology strategies and almost no people strategy. They're asking, "How fast can we deploy AI?" when they should be asking, "Who's left standing when we do?" They're chasing efficiency gains that create catastrophic risk. They're celebrating perfect uptime while their institutional knowledge walks out the door. They're treating the biggest transformation in manufacturing history as if it's just another equipment upgrade.

And they're running out of time to fix it.

What This Book Will Show You

This book isn't about resisting automation. It's about deploying it intelligently, with clear eyes about what technology can and cannot do.

You'll see how judgment erodes in successful operations – not through neglect, but through reasonable decisions that feel responsible at the time. You'll learn how to read your system the way it actually operates, not the way org charts say it should. And you'll get a framework for building what I call a Smart Plant™: an operation where technology and human capability amplify each other instead of competing.

Some manufacturers are already building this. They're not rejecting technology. They're designing systems that preserve the judgment they'll need when algorithms encounter reality's surprises. They're asking better questions. Making smarter investments. Building competitive advantages that compound for decades.

This book is your invitation to join them. Because the manufacturers who figure this out first won't just survive the next twenty years. They will dominate their markets.

In the spring of 2030, a notification arrives at 6:47 a.m. on the phone of an operations executive who believes her plant is performing perfectly. She's about to discover she was catastrophically wrong. Not because of what the metrics show. But because of what they hide.

"We have a talent pool that is transitioning out, and there's no backstop. The question is: How do you take someone with decades of experience and transfer that knowledge to someone who has no skill at all?"

— Ivan Madera, Founder, Morph 3D

CHAPTER ONE:
The Failure No One Saw Coming

Spring 2030

The notification hit Denise Tarlow's phone at 6:47 a.m.: OAKRIDGE INDUSTRIES: 500 CONSECUTIVE DAYS OF ZERO DOWNTIME.

She stared at it longer than she should have. Five hundred days. A number that would look perfect in the quarterly board deck. A number that should have made her proud.

Instead, it made her uneasy in a way she couldn't name.

As Vice President of Operations, she'd been around long enough to know that perfection in manufacturing is always temporary. But this wasn't the usual anxiety about equipment wear or supply hiccups. This felt different. Because Denise had learned something most executives discover too late: When manufacturing looks this perfect, you're either bulletproof or one surprise away from discovering you never were.

By 7:15 a.m., coffee in hand, she was walking the main aisle of the plant, watching the day shift settle in. On the surface, it looked like a masterclass in efficiency. Collaborative robots moved with precision. Conveyors flowed smoothly. Automated guided vehicles kept a perfect pace. The plant was clean. The noise level was minimal. Everything hummed with machine confidence.

But something was off. Not a breakdown. An absence. There used to be more noise, but not mechanical noise. Human noise. Conversations. Coaching.

Real-time adjustments made face-to-face. Small problems surfaced and were solved on the fly.

Now, interactions were filtered through screens. People watched dashboards. Alerts appeared. Suggestions were followed. Work moved on. The plant ran efficiently, but it no longer felt alert.

Marta Hensley sat in her usual spot in the breakroom, the one by the window. She'd claimed it years ago. At fifty-eight, Marta had spent most of her adult life at Oakridge, long enough to sense a problem before the data ever did.

Her role had shifted over the years. Less hands-on with equipment. More time monitoring systems that were supposed to do that for her. What disappeared wasn't the work.

It was the conversations.

Three years ago, training happened in real time. A new hire would shadow her for weeks, not just learning procedures but also absorbing her instincts. When to pause. What hesitation felt like. How to read the rhythm of a line beyond what any sensor could capture. Now, training had become a series of modules completed at a computer terminal.

Devin Rousseau, nine weeks into his employment, still looked to Marta for insight when he could. He hadn't had the curiosity knocked out of him yet. He wanted to understand, not just react. That morning, they stood near Line 4, watching one of the robotic cells cycle endlessly.

"It never stops," Devin said, a mix of admiration and unease in his voice. "Not even for a second."

Marta didn't watch. She listened. After a beat, she frowned. "Something doesn't sound right."

Devin checked the screen mounted at eye level, the one that showed real-time status for every system on the line. All green. No alerts. No deviations. "What do you hear?"

"A bearing's starting to go," she said. "Could last hours. Maybe days."

"Should we log it?"

"The system says everything's fine," she replied, not angry, just stating a fact.

Devin hesitated. The system was sophisticated. Predictive maintenance algorithms analyzed vibration patterns, thermal signatures, and acoustic anomalies. If there were a problem, the sensors would flag it. But Marta had heard something.

"You want me to message my supervisor anyway?" he asked.

Marta shrugged. "You can try."

He did. The response came quickly. "Dashboard shows normal. Focus on your station."

Devin looked at Marta. She didn't say anything. She didn't need to. When the data is green, concern doesn't matter.

> **"My numbers say we're fine, yours say we're not; that's a culture problem."**
>
> *— Alex Ladd, CEO, MindStream Analytics*

Two hours and seventeen minutes later, Line 4 failed. And it didn't fail quietly. The bearing didn't just wear out. It came apart. Debris moved through the system before anyone could stop it. The robots, doing exactly what they were programmed to do, kept running, pushing damaged material forward until the safety system finally shut everything down.

Oakridge, which just hours earlier had celebrated 500 days of perfect uptime, now sat completely idle. For six days.

The Aftermath

The system adjusted immediately. Alerts triggered. Causes were categorized. Workflow steps assigned. On paper, the response looked quick and orderly. Understanding what happened took longer.

Predictive maintenance logs showed nothing unusual. Sensor data stayed within expected ranges. Algorithms ran exactly as designed. When engineers pulled the bearing apart in the forensics lab, they found a metallurgical defect that had existed since installation. An anomaly too subtle for sensors to detect. Obvious to anyone with thirty years of experience who knew how to listen.

The direct costs were easy to tally: $340,000 in repairs, $180,000 in expedited parts shipping, $95,000 in customer penalties for delayed orders. The indirect costs crept in more slowly.

Customer confidence took a hit. Two major accounts requested backup supplier qualification. Timelines slipped across three product lines. The postmortem report grew thinner each time it was revisited because the root cause became harder to articulate as more data was examined.

Then something small surfaced during a leadership review, something that changed how Denise saw the entire event. Almost as an afterthought, someone mentioned that a junior operator had flagged a concern that morning. Devin.

He hadn't shut down the line or made a scene. He followed his training. He noticed something that didn't feel right and messaged his supervisor. The response came quickly: *Dashboard shows normal. Focus on your station.* The data was green. No indicators flagged. Nothing actionable. So nothing happened.

No one scolded Devin. No one told him he was wrong. His concern just didn't go anywhere. In a system designed around data certainty, instinct didn't register. Two hours later, the line crashed. When Denise replayed the sequence in her mind, one realization landed harder than the rest. This failure didn't happen because someone ignored a clear warning. It happened because people had been conditioned not to trust their own judgment.

At a limestone quarry I toured, leaders were explicit that despite advanced sensors and automation, the operation still depended on operators who could feel when something wasn’t right before any system flagged it. Technology handled what it could see. Judgment existed to catch what it couldn’t, but only because the work still required people to act on early unease.

In the Oakridge case, the system didn't silence Devin. Silencing him wasn't necessary. Making him irrelevant was enough. And once judgment stops being valued, the decline isn't loud – the disappearance is silent, which is why no one notices until the capability is gone. Oakridge didn't fail because the technology was flawed. The data showed nothing wrong. That was the problem.

Oakridge failed because the last place where human judgment could have changed the outcome had been systematically engineered out of the operation. If you're reading this and thinking, *That wouldn't happen here,* ask yourself one question: *When was the last time someone in your operation stopped the line because something felt wrong, even though the dashboard was green?*

If you can't remember, you're not different from Oakridge. You're just earlier in the same timeline.

"When it's 2:00 a.m., and Mikey has a machine down, has a line down…operations is down and breathing heavy on Mikey. If Mikey doesn't have that knowledge, that gap is magnified."

— Derek Crager, Founder, Practical AI

CHAPTER TWO:
Three Years Earlier

The training room at Oakridge was loud. Not with machinery, but with argument. Marta stood near the plasma table, watching a new hire, a young woman named Kendra, attempt to diagnose why a cut had gone wrong. The part was scrap now, a $200 mistake that would never leave the building.

"Walk me through it," Marta said.

Kendra recited the procedure exactly as she'd been taught. Feed rate, torch height, gas pressure – all within specification.

"So why did it fail?" Marta asked.

"I don't know. I followed the settings."

"Look at the edge." Marta handed her the ruined part. "What do you see?"

Kendra examined the piece. "It's rough. Inconsistent."

"Why?"

"Maybe the torch height drifted?"

"The system would've caught that. Try again."

Kendra turned the part over in her hands, frustrated. This wasn't how training was supposed to work. She'd completed the modules. She knew the specifications. Everything should have been straightforward.

"I don't know," she finally admitted.

Marta picked up a different piece of metal, one that had been sitting on the worktable. "Feel this."

> **"The next generation needs to learn the granular ground-level aspects before they let tech run everything."**
>
> *— Lee Rector, CEO, LaborAI*

Kendra took it, uncertain what she was supposed to notice.

"Temperature," Marta said. "Material came in cold this morning. The delivery truck broke down overnight and sat in 15-degree weather for six hours before it got moving again. System doesn't know that. Sensors check the table, not the stock. Cold material cuts differently. You have to adjust."

"But that's not in the procedure."

"No," Marta said. "It's not."

The Conversation That Didn't Happen

In early 2028, one of Oakridge's most experienced maintenance leads, Frank Delgado, retired after thirty-four years of service. He had worked at the plant since it opened, back when everything was manual, back in the days when you learned by breaking things and fixing them under pressure.

HR organized a small retirement gathering. People shared stories. Someone joked that Frank could diagnose a motor problem just by putting his hand on the housing. It was true. Frank could. No one talked about whether anyone else at Oakridge could do the same thing. Or whether the work, as it was currently designed, would ever teach someone to be able to do what Frank could do.

Frank left on a Friday. By Monday, his tasks had been redistributed across three people. The system absorbed his departure without friction. Six months later, a motor on Line 3 began showing signs of wear. The predictive system flagged it as not urgent but something to watch. The recommendation was to replace the line during the next scheduled maintenance window, which was three weeks out.

Frank would have pulled it immediately. Not because the data was wrong, but because he'd seen this specific failure mode before. The motor would degrade slowly, then fail catastrophically, usually taking out the coupling and damaging the gearbox in the process. But Frank was gone. And the system, trained on historical data that didn't include Frank's judgment, saw no reason to accelerate the timeline.

Three weeks later, during the scheduled replacement, they discovered the coupling had already started to fracture. Another week and it would have failed exactly the way Frank would have expected.

No one connected the dots. The system had flagged the issue. The part was replaced. Everything worked as designed. What didn't get captured was how close they'd come to a much more expensive failure, and how that risk had once been managed by judgment; the organization no longer systematically developed.

By mid-2028, Oakridge's training program had been "modernized." Instead of spending three weeks shadowing experienced operators, new hires now completed a two-week onboarding process: five days of classroom instruction, five days of simulator training, and four days of supervised floor time. The new program moved faster. More consistent. Easier to scale. And graduates emerged skilled at following procedures but unprepared for the moments that demanded something different.

Kendra, the woman who'd learned about cold material from Marta, had since been promoted to a training role herself. She tried to replicate the kind of teaching she'd received – contextual, situational, built on real problems. But the new system didn't leave room for that. Training was modular. Competency was measured by task completion, not judgment development. She could teach people what to do. She couldn't teach them how to think.

After six months, she left for a competitor. In her exit interview, she said she "didn't feel like she was adding value anymore."

The Turning Point

In late 2028, Oakridge installed an AI system to run its production line. A significant investment – $1.2M – but the business case was solid. The system used machine learning to analyze production data in real time, adjusting parameters dynamically to maximize throughput while maintaining quality.

The results were immediate. Output increased by 11% in the first quarter. Scrap rates dropped. Energy consumption fell. The system learned faster than any human could, finding efficiencies that weren't obvious even to experienced operators.

Denise presented the results to the board in March 2029. The numbers were the best she'd ever delivered. What the data didn't show was how the AI system had noiselessly shifted authority. Operators still had override capability. Technically, they could ignore the system's recommendations. But in practice, overriding rarely made sense. The AI was almost always right. Questioning the recommendations felt like wasting time, or worse, like admitting you didn't understand the math behind the decision. So people stopped questioning them.

They didn't stop caring. They didn't stop paying attention. They just learned that their role was to monitor, not interpret. To confirm, not decide. The system had become the authority. And authority, once ceded, is difficult to reclaim.

By the spring of 2030, Oakridge looked like a model of operational excellence. Metrics held strong across the board. Technology investments had started paying off. Efficiency gains compounded quarter over quarter. Leadership felt confident.

Marta still showed up every day, still punched in, but her role had become almost entirely supervisory. She watched screens more than she watched equipment. She trusted the data because the data rarely lied.

Devin had joined a few months earlier, part of a new cohort of hires who had never known manufacturing without AI. To him, the system wasn't an addition to the work. The system did the work. The training room had grown quieter. Fewer arguments. Fewer detours into context and nuance. Faster throughput.

Everything hummed along efficiently. Until the morning of Thursday, March 28, 2030, when Marta heard something the system couldn't, and no one knew how to act on it.

Here's what you need to ask yourself: How many Martas do you still have? And when they hear something the system doesn't, does anyone listen before it's too late?

"The danger is when there's no human thought.
You miss something of critical importance."

— *Ivan Madera, Founder, Morph 3D*

CHAPTER THREE:

The AI Difference: Why This Time Is Not Like Before

Manufacturing has survived waves of automation before. Mechanization eliminated manual labor. Electrification brought power to every workstation. Computerization enabled precision at scale. Each transition reshaped the workforce, and each time, leaders worried that machines would make people obsolete.

They never did. People adapted. New skills emerged. The work changed, but it remained fundamentally human in critical ways. What's happening now isn't another chapter in the automation playbook. It's a different book entirely.

AI doesn't just execute tasks faster; it replaces the cognitive authority that once separated humans from machines. It observes your patterns. It learns your exceptions. It predicts your failures before you feel the drift. And increasingly, it makes the call before you even know there's a choice to make.

The robots didn't take your jobs. They took something harder to replace: the organizational expectation that humans should think. And most leaders approved that trade without realizing they'd made it.

That's why this transition feels different to the leaders I've interviewed. It's not about physical labor anymore. It's about cognitive authority. And cognitive authority, once transferred to a system, doesn't come back easily.

And here's the part most of you don't want to admit: You are accelerating the erosion. You bought the AI because the vendor promised it would "augment human capability." But you deployed it to replace human decisions. You said people matter, then you designed systems that made them optional. Don't tell me you value judgment when every workflow you've approved in the last two years has been engineered to eliminate it.

> **"AI knows all the how-to stuff. What people pay for is not what you know; it's how you think."**
>
> — *Randy Gage, Author and Business Strategist*

What Makes AI Different

Previous waves of automation replaced physical labor. You swapped a person doing repetitive work for a machine that could perform faster and longer. What's happening now is fundamentally different. AI doesn't just execute tasks; it observes, predicts, recommends, and increasingly, decides. It operates at machine speed in domains that used to require human judgment, the kind of expertise that took years to develop.

When a system can sense a quality deviation before a person notices it, diagnose a maintenance need before a technician hears it, and rewrite a schedule before a planner reviews it, the nature of work changes completely. The human role shifts from doing to monitoring, from interpreting to acknowledging, from deciding to confirming what the system has already determined.

That shift happens gradually, one improvement at a time, until the day something breaks in a way the system wasn't trained to recognize, and there's no one left who knows how to respond.

Modern machine learning systems do more than execute; they interpret. A predictive maintenance algorithm identifies patterns in vibration, temperature, and acoustic data that indicate a bearing will fail three weeks from now, well before any sensor crosses a threshold.

Quality inspection systems identify subtle surface-finish defects that correlate with downstream failures, even when those defects fall within specification. Adaptive production systems dynamically adjust parameters across dozens of variables to maximize throughput while maintaining quality, learning from each run rather than following a fixed schedule.

These systems don't replace muscle. They replace judgment. In manufacturing operations I've toured, leaders describe sensors as necessary but incomplete. Operators still listen for tone changes, vibration patterns, or timing shifts that models can't yet recognize. The concern isn't whether AI works; everyone knows it does. The concern is whether the organization still expects humans to notice what the model hasn't learned yet. When judgment moves from people to algorithms, the nature of work changes completely.

The Speed Problem

AI operates at machine speed. When a quality issue appears, an AI system can detect it, classify it, trace it back to the root cause, and adjust parameters across the entire line, all in the time it takes a human to notice something looks off. That speed is valuable. It prevents defects from propagating. It reduces waste. It protects the margin.

But it also creates a new problem: By the time a human could intervene, the system has already made its decision. This isn't a flaw. It's the design. Speed is the advantage. But speed compresses the window in which human judgment can matter. If intervention requires proof, and proof requires time, and time allows the system to commit the organization to a path, then judgment arrives too late.

At Oakridge, Marta heard the bearing failing. But by the time she could have escalated her concern, gathered evidence, and convinced someone to act, the system had already determined that everything was fine. The system wasn't wrong at that moment. The sensors genuinely showed normal operation. The failure was progressing too slowly for algorithms trained on historical data to recognize.

But Marta knew. Thirty years of experience told her something was wrong. The problem wasn't that the system failed to detect the issue. The problem was that the organization had been designed to trust the system more than it trusted Marta.

When systems move faster than humans can interpret and when authority resides within the system, judgment becomes structurally irrelevant.

The Confidence Problem

Earlier generations of automation were limited by programming. A robot did exactly what it was told to do and nothing more. If conditions changed, it kept executing the same instructions until a human intervened. The limitations were visible, which meant the role of people remained visible as well.

AI is different. It adapts. It learns. It improves with experience. And that adaptability creates a dangerous confidence.

Leaders watch the system get better. They see it catching errors and anomalies that people miss. They see it processing volumes of data that no human could reasonably track. Performance improves. Variability shrinks. Decisions come faster. Of course, they trust the technology. They invested serious money in it, and this is exactly what it was supposed to do. If it didn't work this well, they would have a very different problem on their hands.

But that confidence carries a hidden tradeoff.

AI performs with certainty because it has been trained on historical data. It knows what normal looks like based on the past. What it cannot do? Recognize when the present stops playing by the past's rules. Your system is driving 60 mph using yesterday's map. And you're in the passenger seat, trusting it to notice when the road disappears.

When novelty appears, a supplier slightly changes a material formulation, a new failure mode emerges, or a geopolitical event disrupts logistics overnight, the system doesn't recognize that it is suddenly operating in unfamiliar territory. It processes the situation using old assumptions. Sometimes that's enough. Other times, it leads the organization forward with certainty when caution would be more appropriate.

Human judgment exists for precisely these moments. To notice when something doesn't quite fit. To hesitate when confidence feels premature. To act before the pattern is fully understood. But when an organization has been designed to defer to system certainty, there is no longer a place for human judgment in the decision.

At Oakridge, the predictive maintenance system generated high-confidence outputs because it had been trained on years of bearing data. It understood normal degradation. It knew when to flag concerns based on patterns it recognized. What it didn't know was that this particular bearing contained a metallurgical defect that produced an entirely different failure signature.

The system wasn't reckless. It wasn't malfunctioning. It was operating exactly as designed. The failure occurred when its confidence became the organization's confidence. When Marta's instinct conflicted with the dashboard, the dashboard prevailed.

When organizations treat AI-generated certainty as truth rather than as one input among many, they lose the ability to act in moments of uncertainty. And uncertainty is where judgment has always mattered most.

The Learning Problem

If confidence is the first shift, learning is where the real damage accelerates. Traditional automation didn't learn. If you wanted a robot to do something new, you programmed it. AI learns continuously. Every decision it makes feeds back into the model. Every outcome refines its understanding. Over time, it gets better at what it has been trained to do. That's powerful. It's also why AI accelerates the erosion of judgment faster than any previous form of automation.

Here's how it works: An AI system makes a recommendation. A human confirms it. The outcome is positive. The system logs the recommendation as validated and increases the probability weighting for that decision pattern. The next time a similar situation arises, the system generates the same recommendation with a higher certainty score. Humans, having seen the system succeed, are more inclined to defer.

Over time, the human stops evaluating the recommendation and starts simply confirming it. The system learns that its recommendations are almost always accepted, which further reinforces its confidence. Eventually, the human role becomes procedural. The system decides. The human acknowledges. This occurs because the system is genuinely good at what it does. Most of the time, deferring to its analysis is the right move. But "most of the time" isn't the same as "always."

The system learns from success. It doesn't learn from near-misses it never detected. It doesn't learn from failures that were prevented by human intervention before they became visible in the data. Over time, the organization becomes excellent at handling what the system has been trained for, and increasingly vulnerable when facing what it hasn't.

At Oakridge, the AI-powered production system learned from every shift. It got better at maximizing throughput. It got better at reducing scrap. It got better at balancing dozens of variables simultaneously. What it didn't learn was how to recognize when those improvements were pushing the line into a fragile state, running too close to tolerance, deferring maintenance too aggressively, or relying too heavily on a single supplier.

Those risks weren't visible in the data. They lived in judgment. And judgment, no longer required to make decisions, had stopped developing. That's the learning trap. AI gets smarter about what it knows. Organizations get weaker at everything else.

"The risk is getting too lazy, too complacent. AI is subordinate to you. It's not there to do everything for you."

— Eknauth Persaud, CEO, RentMySoftware.com

Why Leaders Miss This

When I ask manufacturing leaders about AI, the conversation almost always starts the same way. They want to know what it can do, how accurate it is, how fast the return will be, and how it compares to what their competitors are using.

Those questions are the easiest ones to ask when you feel pressure to act fast before you fully understand. No one wants to be the plant that appears outdated, or the leadership team that can't attract the next generation of talent because their technology seems stuck in another decade. So leaders move forward, often with a sense that they'll figure it out as they go.

Here's the problem. That mindset treats AI as a tool you install instead of a system you introduce. And tools can be patched. Systems cannot.

A band-aid approach asks, "What problem does this solve right now?"

A system approach asks, "What behaviors does this reinforce over time?"Every interaction with an AI system teaches people something. Not in a formal training sense, but through repetition. When the system makes a recommendation and it's right, people learn to trust it. When it flags an issue and prevents a failure, people learn it's reliable. When it consistently outperforms human judgment in narrow tasks, people learn to defer. The system is doing exactly what it was designed to do.

But taken together, those moments reshape roles. People stop interpreting and start monitoring. They stop questioning and start confirming. They stop deciding and start executing. Over time, the organization trains itself into compliance with system output – even when no one ever said that was the goal.

At Oakridge, no one declared that human judgment no longer mattered. There was no announcement, no policy change, no memo. Over three years and hundreds of small interactions, the message was delivered anyway. Marta heard the bearing. Devin raised the concern. Both had learned through experience that when instinct conflicted with the system, the system carried more weight.

That's not a technology failure. That's a design failure.

When leaders adopt AI to keep up, check a box, or signal modernity, they often miss what they're actually building. Not just new capability, but new habits. Not just faster decisions, but narrower ones. And by the time the system encounters something it doesn't recognize, there may be no one left who is practiced at acting without its permission.

This is the shift leaders need to own. AI is never just about what it can do. It's about what it conditions your people to *stop* doing.

The Pace of Change

The final difference is pace. Previous automation arrived slowly. Organizations had time to adapt. Training programs evolved. New roles emerged. People learned how to work alongside machines over years, sometimes decades. AI is moving faster.

Systems that took months to implement five years ago now deploy in weeks. Capabilities that required custom development are now available as cloud services. Tools that cost millions are now accessible at scale. That acceleration is changing how organizations learn.

When automation arrived slowly, knowledge transfer happened naturally. Experienced people worked alongside new technology long enough to figure out how to integrate it into their judgment. They learned what the system could do, what it couldn't, and when to trust it. When AI arrives quickly, that integration doesn't happen. The system is deployed. It works. People adapt to it. But they don't develop judgment around it, because there isn't enough time.

At Oakridge, the production AI went from pilot to full deployment in four months. It was a success story internally. Fast implementation. Immediate results. Minimal disruption. What didn't happen was the slow, messy process of people learning how to work with the system. How to question its recommendations. How to recognize when it was chasing the wrong outcome. How to intervene when its confidence exceeded its competence. That learning takes time. And time is the one thing rapid deployment doesn't provide.

What This Means

AI is not just another tool. It's a different category of technology, one that operates in the space where human judgment used to be the only option. That doesn't make it dangerous by default. It makes it consequential.

Organizations that treat AI deployment as a technology project will automate their operations while quietly hollowing out judgment. Organizations that understand AI as a shift in cognitive authority – who decides, when, and based on what – have the opportunity to build systems that are both intelligent and resilient. The difference between those two paths isn't the technology. It's leadership. Not leadership that resists AI, but leadership that governs it.

That governance starts with a question most organizations never ask: What is our technology teaching our people?

At Oakridge, the answer didn't surface until it was too late. Over time, the system taught people that judgment was optional. When it finally mattered most, judgment wasn't available.

Before you turn the page, ask yourself this: In your operation today, when the system is confident, what are your people expected to do? Confirm it or challenge it? Follow it or interpret it? Pause it or stay out of its way?

And here's the more difficult question: If someone challenged a confident system recommendation today and turned out to be wrong, what would happen to them? If your honest answer involves words like "explaining themselves" or "justifying the disruption," you've already taught them not to bother.

Because the way your organization handles confident systems today is training people for tomorrow. If your people are expected to confirm instead of question, follow instead of interpret, and stay quiet when the data looks clean, that's exactly what they'll do when the situation is no longer clear.

And at some point, it won't be clear. Conditions will change. Context will shift. The system will face something it has never seen before. When that happens, the only decisions available to you will be the ones your people have practiced and been permitted to make.

What Is Actually at Risk

Before we examine how these risks hide behind perfect metrics, let's be clear about what is actually at stake.

If you run or own a manufacturing facility, here's the part the dashboards don't show and the part that threatens your business long before AI replaces a single job:

You are losing your people pipeline. Your future supervisors, managers, and technical experts aren't being developed because development requires time, mentorship, and human connection – three elements AI-driven systems dismantle in the name of speed.

Your bench strength is evaporating. Retirements are accelerating. New hires typically stay for a mere fourteen to eighteen months. Your institutional knowledge is aging out faster than you can digitize it.

Your automation investment is at risk. You can buy robots, but you cannot buy a workforce ready to run them. Without the support of a workforce with the right skills, automation becomes brittle, fragile, and expensive.

You are competing against giants you cannot outspend. Companies like Amazon, Bosch, and Toyota will pull talent from you unless you create a culture they cannot replicate.

AI gives a false sense of certainty. Things appear stable right up until they fail.

Most owners don't fear robots. They fear uncertainty. Missed signals. Problems they didn't see coming. Without a people strategy in place, leaders allow AI to reinforce the illusion that everything is under control.

Calm doesn't mean resilient.

"Sometimes all these issues of high absenteeism or communication breakdown are symptoms of a very deep-rooted issue. And we need to be okay with wanting to walk through that and understanding that there's going to be growth beyond that."

— *Marilyn Rosa-Green, Founder/CEO, Marilyn Rosa-Green Consulting Solutions*

CHAPTER FOUR:

When Perfect Metrics Hide Fragile Systems

The morning Line 4 failed, every metric at Oakridge was green.
Throughput: 102% of target.
Quality: 99.4% first-pass yield.
Uptime: 500 consecutive days without unplanned downtime.
Safety: 127 days since last recordable incident.
Efficiency: Operating at 94% of theoretical maximum.

By every leadership measure, Oakridge was performing exceptionally well. And by every measure that mattered to resilience – the organization's ability to handle surprise – Oakridge was becoming catastrophically weak. They were measuring the wrong things.

> **"Most companies don't post the productivity of their people. They post productivity of production."**
>
> *— Karla Trotman,*
> *President & CEO,*
> *Trotman Manufacturing*

What Metrics Actually Measure

Metrics answer one question: How well is the system performing against known expectations? That's useful, until it isn't. Because metrics can't tell you whether your operation can handle what it hasn't seen before. You're measuring execution when you should be measuring readiness. And readiness doesn't show up on a dashboard until it's too late.

Resilience isn't visible in performance data. It appears in how an organization responds when assumptions break. When a supplier fails. When a material behaves unexpectedly. When a machine degrades in a way the model wasn't trained to recognize. Performance metrics measure execution. Resilience depends on judgment. And judgment doesn't show up on a dashboard.

The Dashboard Trap

Dashboards are seductive. They compress complexity into clarity. They turn ambiguity into color. Green means good. Yellow means watch. Red means act. We've been trained to trust those signals since childhood. They allow leaders to make quick, confident decisions. They also create a blind spot.

When the dashboard is green, the concern feels unfounded. When all indicators point to success, questioning the system feels like resistance. When metrics are strong, an instinct that contradicts them gets reclassified as noise.

At Oakridge, Denise walked the floor every morning. She'd built that habit years ago, back when metrics were simpler and problems announced themselves through visible signs: a machine running rough, an operator looking uncertain, or material piling up in the wrong place.

But as the dashboards became more sophisticated, the walks became less diagnostic. She was no longer looking for problems. She was confirming what the data had already shown. When everything looks good on the dashboard, the floor walk becomes a formality. And formalities miss surprises.

The Silence That Looks Like Health

One of the most common things I hear from manufacturing leaders sounds like this:

"Our plant is running better than it ever has. So why does it feel more fragile than ever?"

They're not talking about output. The numbers look good. Scrap is down. Safety metrics are solid. On paper, everything is working. What's keeping them up at night is people. Finding them. Keeping them. Replacing decades of experience before it walks out the door for the last time.

They know their veteran employees are retiring faster than knowledge can be transferred. And they know a knowledge transfer plan isn't the same as transferring twenty years of instinct built from listening to that machine. They see newer hires who are comfortable with screens and dashboards but don't yet have the instincts to diagnose a piece of equipment that's been running since the '80s.

Some can manage the software, but they've never turned a wrench on the older machines that still carry a big part of the operation. Now layer in new automation and AI on top of that, and the gap gets wider, not smaller.

What these leaders are sensing is the difference between performance and capability.

Performance is what the system delivers today. Capability is what the organization can handle when something changes tomorrow. Strong performance can hide declining capability for a long time, especially when technology is compensating for what people no longer know.

I saw this clearly at one quarry operation. By traditional measures, the plant looked stable. Turnover was low. Output was strong. Safety numbers were clean. But the workforce underneath those metrics was changing fast. In just three years, employees with more than thirty years of experience dropped from roughly 40% of the workforce to closer to 15%. Most of the new hires had been there fewer than three years.

Nothing broke. Production kept moving. But the organization crossed a threshold where judgment transfer could no longer keep pace with loss, the kind of erosion dashboards aren't designed to detect.

Everything keeps working. Until it doesn't. And when that moment comes, leaders discover they no longer have the margin they thought they did.

What Gets Lost First

When operations revolve around metrics, certain capabilities erode predictably.

Early pattern recognition disappears. When the system flags problems, people stop looking for early signals. Why spend energy noticing subtle changes when the dashboard alerts you if something matters?

Informal knowledge transfer stops. When training is modularized and measured by task completion, the messy, inefficient process of teaching someone how to think gets deprioritized. It doesn't appear as a competency metric, so it isn't protected.

Intervention confidence weakens. When people are conditioned to wait for confirmation before acting, they lose the muscle of moving on instinct. Over time, even experienced operators start questioning what they notice if the data doesn't confirm their read.

Cross-functional sensemaking fades. When each department optimizes its own metrics, informal conversations used to surface problems early: production talking to maintenance, quality talking to engineering, become less frequent.

Everything is legible within its own domain, but the handoffs go dark.

These are the capabilities that matter most when something unexpected happens. At Oakridge, all four had eroded over three years. The metrics never flagged the decline because metrics don't measure capability. They measure output. And output stayed strong right up until the operation collapsed.

The Optimization Paradox

The better a system performs, the more dangerous certain kinds of optimization become. When metrics are consistently green, leadership naturally asks: Can we go further? Can we tighten tolerances? Can we reduce inventory? Can we defer maintenance? Can we shorten training?

Each optimization creates value. Each efficiency gains compounds. But together, they reduce the margin. Margin is what gives people time to notice, interpret, and respond before a small problem becomes a large one. High-performing systems optimize margin away because margin looks like waste.

Extra inventory? Inefficient. Redundant sensors? Unnecessary. Manual checks that duplicate automated ones? Obsolete. Training time beyond minimum competency? Expensive. Every efficiency initiative makes the system faster, leaner, and more predictable. And more fragile.

At Oakridge, the margin had been disappearing for years. Not recklessly, but responsibly. Data justified every decision. Every optimization delivered measurable value.

What didn't get measured was how close the plant was operating to the edge of its capability. When the bearing failed, there was no buffer. No redundancy. No one who could intervene before the system committed. The improvements that made Oakridge efficient is what made it brittle.

Why Leaders Can't See It

The optimization paradox is hard for leaders to see for one simple reason: things are working. I've never had a leader call me and say, "Lisa, our metrics are incredible, and we're way ahead of plan. Can you come help us?" They call when something breaks. When a customer escalates. When a key person leaves. When a failure exposes how thin the margin really was.

Strong metrics create confidence. Confidence reduces scrutiny. Reduced scrutiny means small erosions go unnoticed. By the time a leader realizes margin is gone, it's usually because something has already failed. And by then, the conversation shifts to recovery, not prevention.

What makes this especially hard is that the people who can see the erosion happening – the Martas, the experienced operators, the long-tenured maintenance techs – are the same people whose authority has diminished beneath awareness. But they've learned that concern without data doesn't count, so they stop raising their hand.

At Oakridge, Marta had seen the margin disappearing for years. Fewer experienced people on the floor. Less time for informal coaching. More reliance on systems that couldn't interpret context. She raised the concern once in a skip-level meeting with Denise.

"It feels like we're running leaner than we used to," she'd said.

Denise had looked at the dashboard. Productivity was up. Efficiency was strong. Turnover was lower than the previous year.

"Can you be more specific?" she asked.

Marta couldn't. She didn't know how to distill her unease into one concise concern. It was a sense, built on years of experience, that the plant was becoming more precarious even as it was becoming more efficient. Without data to support it, the concern faded. The people who saw it couldn't prove it. And the people who could act on it couldn't see it.

The Measurement Trap

Organizations measure what they value. But they also come to value what they measure. Once a metric becomes the definition of success, behavior aligns around it. People are not necessarily gaming the system; they're doing their jobs.

If uptime is the primary metric, people optimize for uptime. They defer maintenance to avoid downtime. They avoid stopping the line unless necessary. They design workflows that minimize interruption. All of that makes sense if uptime is the goal. But if uptime becomes the only goal, judgment erodes.

Because judgment often requires stopping. Pausing to investigate. Slowing down when something doesn't feel right, even if the data says everything is fine. When stopping is penalized and continuity is rewarded, judgment retreats.

At Oakridge, the 500-day uptime streak wasn't just a milestone. It was an identity. Every morning, the number ticked higher. Every shift handoff included it. Every leadership meeting referenced it. No one explicitly said, "Don't stop the line." But the message was clear. Stopping was failure. Continuity was success.

When Devin messaged his supervisor about Marta's concern, the supervisor faced a choice: trust instinct and risk interrupting the streak or trust the data and keep things moving. The data was green. The streak was alive. The choice was obvious.

That's the measurement trap. When metrics define reality, anything that contradicts them becomes suspect.

What Should Be Measured Instead

If metrics can hide fragility, what should organizations measure? The answer isn't to stop measuring performance. It's to add measures of capability. Performance metrics tell you how well the system is executing today. Capability metrics tell you whether your organization can handle what it hasn't seen before. Both matter. But most plants measure only the first.

Your dashboards are lying to you. Not because the data is wrong, but because they're measuring the wrong thing. Performance metrics tell you how well you executed yesterday. They can't tell you whether you'll survive tomorrow. What follows are seven capability metrics that most manufacturers never track, and every one of them predicts failure better than your current KPIs. These aren't nice-to-haves. These are early warning systems. And if you're not measuring them, you're flying blind.

1. Signal-to-Action Time

Track how long it takes between when someone first notices something unusual and when action is taken. Not when an alert fires, but when a *human* notices. If this window shrinks, your organization is getting faster at responding to early warnings. If it's growing, people are hesitating longer before acting, which usually means they've learned that early signals don't count without data confirmation.

At Oakridge, Marta's concern never became actionable. It went from observation to dismissal without ever triggering an investigation. Years earlier, a similar observation would have prompted immediate attention. The system had no way to measure that shift.

2. Intervention Frequency

Count how often people pause work to investigate something that doesn't show up on the dashboard. A declining intervention rate might indicate maturity, fewer disruptions, and smoother operations. But it often signals something else: people have stopped trusting their instincts enough to slow down momentum.

Interventions that find nothing aren't a waste; they're evidence that judgment is still active. When interventions disappear entirely, it usually means judgment has too.

3. Near-Miss Reporting Rates

Track how many "almost failures" get surfaced each month. If near-miss reports are declining while your operation is scaling, that's not safety improvement; it's silence. People have learned that reporting something that didn't result in an actual failure isn't valued, or worse, invites scrutiny about why they didn't prevent it.

High near-miss reporting is a sign of psychological safety. It means people believe their observations matter before consequences prove them right.

4. Cross-Functional Sensemaking Instances

Measure how often people from different departments collaborate to solve ambiguous problems, issues that don't fit neatly into one domain. Production talking to maintenance about a machine that's "running weird but still in spec." Quality talking to engineering about a defect pattern that's statistically insignificant but feels wrong. Supply chain talking to operations about a vendor that's technically performing but raising concerns.

These conversations don't appear in formal problem-solving logs because they occur before problems are formalized. When they disappear, issues that should have been caught early don't surface until they are crises. If cross-functional collaboration is declining, it usually means departments are optimizing independently, creating blind spots at handoffs where most surprising failures originate.

5. Judgment-to-Procedure Ratio in Training

Measure what percentage of onboarding time is spent on interpretive skill development versus procedural task execution. Take a sample of new hire training hours and categorize them: How much time is spent on "Here's how to do this task" versus "Here's how to think about this process"?

If training is overwhelmingly procedural with minimal interpretive development, you're producing people who can execute but can't adapt. If that ratio has shifted over time, becoming more procedural and less interpretive, you're systematically reducing the organization's capability to develop judgment.

6. Experience Distribution Across Shifts

Track the tenure balance on each crew. Calculate the average years of experience per shift and watch how it changes over time. If you consistently pair inexperienced operators with veterans, knowledge transfer can occur. If shifts are becoming homogeneous –all experienced or all new – you're creating vulnerability. A shift staffed entirely by people with minimal tenure isn't just a coverage issue. It's a judgment gap. When something unusual happens, there's no one who's seen it before.

At Oakridge, the tenure of the night shift had collapsed over several years. The most experienced person on the crew had been there for fewer than two years. The day shift wasn't much better. No metric flagged this as a risk because productivity stayed strong. The erosion was invisible until it mattered.

7. System Override Frequency

Measure how often people question or override system recommendations and whether that rate is increasing or decreasing. This is the most uncomfortable metric on the list, because many leaders instinctively want override frequency to be low. "The system is almost always right, so why would we want people second-guessing it?" But a declining override rate can signal danger. It might mean the system is getting better. Or it might mean people have stopped trusting their judgment enough to challenge it.

The goal isn't high override frequency. It's stable, selective disagreement even as system confidence increases. If your AI is getting smarter and override rates are dropping to near-zero, that's a warning sign. It means authority has fully transferred to the system, and people have learned that challenging it is inefficient or unwelcome. What you want to see is informed disagreement. People override rarely, but when they do, it's because they're noticing something the system can't.

Here's the uncomfortable truth: If you run this diagnostic across your operation right now, at least four of these seven metrics will reveal erosion you didn't know was happening. That's not a criticism. That's the nature of capability loss – it hides inside performance gains. The question is whether you are willing to look.

None of These Metrics Is Perfect

These measures are messy. They're harder to track than uptime or throughput. They don't fit cleanly into existing dashboards. Some of them require qualitative judgment to interpret. That's exactly why they matter.

Capability doesn't show up as a clean number. It shows up in behaviors that are easy to miss. A maintenance tech who still does manual walkthroughs even though sensors cover everything. An operator who logs an observation that doesn't trigger any threshold. A supervisor who takes time to explain why something works, not just how to operate it. These behaviors not only optimize performance today; they build capacity for tomorrow's surprise. Traditional metrics tell you whether you're winning the game you're currently playing. Capability metrics tell you whether you'll still be able to play when the rules change.

At Oakridge, every performance metric was green the morning Line 4 failed. Every single capability metric, had Oakridge been tracking them, would have been signaling erosion for years. The difference is that performance metrics measure outcomes. Capability metrics measure readiness. And readiness determines whether your next surprise becomes a recoverable event or a crisis that reveals how exposed you've become.

The Real Cost

The cost of hidden fragility isn't visible in quarterly results. It appears later, when conditions change and the organization discovers it no longer has the margin it thought it had.

At Oakridge, the direct cost of the Line 4 failure was $615,000. The indirect cost was far higher. Two major automotive customers added backup supplier requirements to their contracts. One of them shifted 30% of its volume within six months. Recruiting became harder. The failure made regional news. Experienced operators at other plants heard about it. Oakridge's reputation as a stable employer took a hit.

Internal confidence eroded. People who had trusted the system started questioning it, but they didn't know what to trust instead. Judgment hadn't been practiced. When they needed it, the ability to act independently wasn't there. Most damaging of all was what the failure revealed about leadership's understanding of their own operation.

Denise had believed the plant was strong. The metrics said it was. The technology said it was. Every indicator pointed to operational excellence. The failure proved the plant was fragile. And fragility, once exposed, is difficult to fix quickly. Because you can't rebuild margin under pressure.

Where This Leaves Leadership

Perfect metrics don't guarantee resilience. They guarantee performance within known conditions. When conditions change, when suppliers fail, when materials behave differently, when equipment degrades in novel ways, performance metrics become lagging indicators. By the time they turn yellow, it's already too late.

Resilience requires something metrics can't capture: the organizational capability to notice, interpret, and act on signals that don't fit the model. That capability is judgment. And judgment, as Oakridge learned, doesn't preserve itself just because performance is strong. It has to be designed for. Protected. Measured differently. Or it disappears behind positive metrics that say everything is fine.

Oakridge's dashboard was green the entire time. Every number said the operation was strong. The question no one asked was the one that mattered: What isn't this measuring?

Here's your homework: Pick one metric you're most proud of. One that's been consistently green. Now ask yourself three questions:

1. What capability am I not measuring that this number depends on?
2. If the people who make this metric possible leave tomorrow, how long would the number stay green?
3. And what would fail first that I'm not currently tracking?

If you can't answer those questions, your dashboard isn't showing you strength. It's showing you borrowed time.

"A lot of businesses have the IIPP [Injury Illness Prevention Program]. It's sitting on their bookshelf somewhere, collecting dust. What they struggle with is implementing the IIPP."

— *Eric Wick, Founder, Safety Team Technologies*

CHAPTER FIVE:

The Decisions That Made This Inevitable

By the time Oakridge shut down Line 4, the system had done exactly what it was built to do: detected what it was programmed to detect, followed its rules, and escalated alerts as designed. The failure was organizational, not technical.

What failed were decisions made long before the line ever went down. Small choices that felt responsible in the moment. Choices that never triggered alarms. Choices that made sense. That's why they were dangerous.

> **"Change is hard. But doing nothing isn't really a great option."**
>
> *— Jason Vanzin, Founder and CEO, Right Hand Technology Group*

Because Oakridge's path to instability wasn't paved with recklessness or neglect. It was paved with reasonable decisions, each optimizing for something measurable, each trading off something that wasn't. This sequence isn't unique to Oakridge. It most often shows up in plants that are disciplined, proud, and performing well.

No One Chose This

No one at Oakridge made an explicit decision to sideline human judgment. There was no meeting where someone declared experience optional. No memo instructing people to trust dashboards over instinct. No strategy that positioned automation as a replacement for thinking.

And yet that's exactly what happened.

This is the gap most leaders don't realize they're stepping into. I've worked with leaders who genuinely care about their people and still end up here. One executive told me, "None of these decisions felt big at the time. We were just trying to keep up."

The training module was shortened to speed up onboarding. A manual check was removed because sensors were more reliable. A senior technician wasn't replaced after retirement because budgets were tight. A supervisor was reassigned because the AI system reduced the need for oversight. Each decision solved a visible problem. Together, they created a problem that no one anticipated.

Decision One: Shortening Training (March 2028)

The conversation started in an HR review meeting.

"Current onboarding takes three weeks," the training manager explained. "But most of that time is shadowing experienced operators. If we moved to a modular-approach – classroom, simulator, and supervised floor time – we could cut that to two weeks and maintain consistency across shifts."

"What's the risk?" Denise asked.

"Minimal. The simulator covers 90% of what people encounter. The supervised floor time handles the rest."

"What about the other 10%?"

"Edge cases. They'll pick it up over time."

It was a reasonable pitch. Three weeks felt long, especially when new hires weren't productive for most of it. The company was growing. Hiring was accelerating. Faster onboarding meant faster productivity. Denise approved the change.

The data never captured what happened during that third week of shadowing. That week wasn't about learning procedures. Experienced operators didn't just teach tasks; they taught skills. They taught attention. What to listen for. What hesitation felt like. When to trust the data and when to trust your gut. That kind of teaching resists standardization. Defies measurement. And when training shortened, the teaching disappeared with it.

Leaders tell me, "We've trained them. They passed the modules. They know the job." And they're right…well, on paper. What's missing is time. Time to struggle a little. Time to ask the dumb question. Time to stand next to someone experienced and hear, "Here's why this doesn't feel right, even though the procedure says it's fine."

When you compress training, you don't just speed up people. You remove the space where judgment is formed. And once that space is gone, it doesn't come back on its own. The new hires who came through the streamlined program were competent. They followed procedures well. They achieved productivity faster. What they didn't develop was judgment. Because the work no longer required it long enough for the habit to form. When someone needed judgment, it wasn't there to deploy.

Decision Two: Removing the Manual Check (June 2028)

Line 4 included a manual quality check as part of the process. After each batch, an operator physically inspected a sample to check for defects that the automated vision system might miss. It was a holdover from earlier days, before the vision system had become as sophisticated as it had.

In June 2028, during a continuous improvement review, someone raised a question: "We've had the vision system running for two years. It's caught 847 defects. The manual check hasn't caught anything it missed. Why are we still doing it?" The data was clear. The manual check was redundant.

"It adds ninety seconds per batch," the operations manager noted. "That's 4.5 hours a week. At our volume, that's real capacity we're leaving on the table."

Denise looked at the numbers. The vision system had a 99.7% detection rate. The manual check added cost without adding value. She approved the removal.

What the data didn't capture was why the manual check had never caught anything the vision system missed. It wasn't because the check was useless. It was because the operators performing the checks had learned over two years that their observations didn't matter whether the vision system said the part was good. The manual check had become a formality long before it was officially removed.

It was the last moment in the process when a human was expected to exercise independent judgment. Once that moment was gone, the process became fully automated in practice, even though people were still technically involved.

Decision Three: Not Replacing Frank (September 2028)

When Frank Delgado retired, the decision not to replace him seemed practical. His responsibilities were split across three roles. The predictive maintenance system had reduced the need for manual diagnostics. From a workload and budget perspective, nothing looked strained.

What wasn't visible was how much risk had been concentrated in a single person.

Frank didn't just fix equipment. He carried context. He remembered which issues were real problems and which ones only appeared to be problems. He knew which workarounds were harmless and which ones created downstream trouble.

That knowledge was never written down because it never needed to be. It lived in the decisions Frank made every day without calling attention to them.

Sometimes, a Frank was the maintenance lead. Sometimes an operator. Sometimes an owner who can walk the floor and notice when something isn't behaving the way it usually does. Every organization has at least one person like this. And most leaders know exactly who it is.

At Oakridge, Frank had seen failure modes that hadn't shown up in years. He knew which components failed early, even when the specs said they shouldn't. He understood how certain machines behaved differently when pushed to the edge of tolerance. None of that appeared in the system – not because the knowledge was ignored, but because the system was never designed to hold that kind of knowledge.

Six months after Frank retired, the motor on Line 3 began showing early signs of wear. The predictive maintenance system flagged it for replacement during the next scheduled window. The motor was changed before it failed catastrophically. Production wasn't interrupted. On the surface, the decision looked validated.

But the outcome masked the exposure.

Frank would have recognized the signature immediately and acted sooner. The system didn't and couldn't, because it had never learned from his decisions. The organization didn't lose Frank the day he retired. It lost the margin he had been quietly providing.

Most leaders don't worry about this because nothing breaks right away. The operation keeps running. The numbers stay strong. The question that lingers is the one leaders rarely say out loud: Who are we relying on right now that we're afraid to lose, and what happens if we do?

That question doesn't show up on a dashboard. But it's one every resilient operation eventually must answer.

Decision Four: Reducing Supervision (January 2029)

By early 2029, the adaptive AI production system had been running for several months. It was working exceptionally well, adjusting parameters in real time, balancing throughput and quality, learning from every shift.

During a leadership review, someone raised an observation: "The system is handling a lot of what supervisors used to do. Real-time problem solving. Parameter adjustments. Exception handling. Are we overstaffed?" The data suggested they might be.

Supervisor interventions declined by 60%. Most shifts ran smoothly without significant human oversight. The system was doing the work.

"What if we moved from six supervisors to four?" the operations manager proposed. "The AI handles the tactical stuff. Supervisors focus on the issues that actually need human judgment."

It was a compelling case. Labor costs would drop. Efficiency would improve. The supervisors who remained would have more strategic roles. Denise approved the change.

What no one anticipated was how removing supervisors would change the dynamics on the floor. Supervisors hadn't just solved problems. They'd been the escalation point for uncertainty. When an operator noticed something unusual, they'd flag it to their supervisor. The supervisor would investigate because the operator's concern warranted attention.

With fewer supervisors and more area to cover, that dynamic changed. Operators still noticed things. But raising a concern now required more justification. Supervisors were busier. The threshold for what counted as worth mentioning went up. Over time, operators learned to self-filter. If the dashboard was green, the concern probably wasn't significant enough to bother the supervisor. Uncertainty stopped surfacing.

When supervisors disappear, uncertainty lingers with nowhere to land. People continue noticing; the system simply makes acting on those observations more difficult.

Decision Five: Trusting the Algorithm (March 2029)

This wasn't a single decision. It was a thousand small ones, made across every shift, every day. When the AI production system recommended a parameter change, operators had the authority to override it. In practice, they rarely did.

Why would they? The system was almost always right. It optimized faster than any human could. It balanced variables that no one could track manually. Questioning it felt inefficient. Following it felt professional. Over time, "the system recommends" became "the system decides." No policy changed. Authority technically remained with the people. But authority that's never exercised isn't authority. It's theater.

"People resist change when there's no 'why.' They've been doing it this way for all these years."

— Lisa Sanderson, Manufacturing Executive, Gleicher Manufacturing Corp.

By the spring of 2030, when Marta heard the bearing failing, the culture at Oakridge had shifted completely. The system was the source of truth. Data was the authority. Instinct was suspect. Devin raised Marta's concern because he was new enough to believe it might matter. His supervisor dismissed it because he'd learned that concerns without data don't count. Everyone followed the process. The process no longer included room for judgment.

Where Responsibility Lives

Let's be uncomfortably honest: you made these decisions. Not "the market." Not "the board." You. You shortened the training because three weeks felt long, and HR said the program could work in two. You removed the manual check because Excel said it was redundant, and no one wants to be the person defending ninety seconds of "wasted" time. Every reasonable choice you celebrated last year is the fragility you're managing today.

Congratulations. You've optimized your way into a corner.

And collectively, those decisions reshaped the organization in ways no single decision-maker could see in the moment. The failure wasn't in the technology or the people. It was in the design of leadership responsibility. Reasonable decisions were made, but no one owned the job of protecting what couldn't be measured while everything else was being enhanced.

Performance improved while capability eroded. The consequences didn't show up during the decision-making. They surfaced later, under pressure, when there was no time to rebuild what had been lost.

Here's your assignment: Open your calendar. Find the last three operational-improvement meetings. Now answer this: In any of those meetings, did anyone ask what capability you were trading away for that efficiency gain? Did anyone ask what judgment you were removing to achieve that cost reduction?

If the answer is no, you're not governing tradeoffs. You're just approving them. And what you're approving is accumulating into a problem that will announce itself when you can least afford it.

That responsibility doesn't belong to HR. It doesn't belong to IT. It doesn't belong to the operators or the algorithms. It belongs to you.

"I've had them come in and say, 'Why is this set up this way?' What do you think it should be? And they change it. And then they have pride in the fact that they change it, and usually it's an upgrade for us."

— *Karla Trotman, President and CEO, Electro Soft Inc.*

CHAPTER SIX:
The Two Paths Diverge

By the time leaders feel the need to respond, most of the direction has already been set. The systems are live. The workflows are established. The roles have been shaped around what technology now handles.

At Oakridge, the conversation after the shutdown focused on prevention. How to make sure the system catches the next issue sooner. How to reduce the chance of another interruption. That question doesn't get asked by the system.

> **"We're losing tribal knowledge as Baby Boomers retire. That's the next supply chain awakening."**
>
> *— John Kevin Koehler, Supply Chain Executive, KK Tool Company*

Two Organizations, One Trajectory

I've worked with plants where the numbers were solid and the leadership team was proud of the progress they had made. What caught my attention wasn't anything I saw on the floor, but what I heard in the conversations. Fewer "what if" questions. Fewer stories about how problems were solved. More confidence in the dashboards and less curiosity about what sat behind them.

That's usually when I know which path an organization is on, even though the leaders don't yet.

Most leaders believe their operation is evolving along a single trajectory:

continuous improvement powered by better tools. They adopt similar technologies. They chase the same efficiency targets. They remove the same sources of variability. From the outside, many plants look identical. The difference shows up in what the work still asks of the people doing it.

I've seen two plants, similar in size and industry, both implementing advanced automation. Both reporting strong performance. Both celebrating efficiency gains. Three years later, one is thriving. The other is struggling with turnover, quality inconsistencies, and a leadership team that can't quite explain why the metrics are strong, but the operation feels fragile.

The technology wasn't the difference. The difference was in what each organization chose to protect while deploying it.

Path One: When Work Stops Demanding Judgment

In some organizations, automation steadily narrows the role. Intervention becomes rare. Questions become unnecessary. Experience matters less than compliance. The work gets quieter. Cleaner. More predictable. People are expected to respond when prompted, not to notice when something is off. Over time, they adapt to what the job requires, which is less thinking. And no one complains about that, until they're asked to think again.

One operations leader put it to me this way: "We used to have operators. Now we have monitors. They watch the screens. They confirm what the system tells them. They're good at their jobs, but those jobs aren't what they used to be."

This path isn't chosen deliberately. It emerges through optimization. Each efficiency gain removes a decision point. Each automation layer reduces the need for interpretation. Each standardization effort narrows the range of acceptable responses. Performance improves. Metrics strengthen. And slowly, the organization becomes less capable of handling anything the system wasn't designed to see.

The uncomfortable question for leaders on this path is simple: If the system went dark tomorrow, who in your operation would still know what to do, and who would be waiting for instructions?

Path Two: When Work Still Teaches

In other organizations, leaders notice when that narrowing begins. They pay attention when roles stop requiring interpretation. They hear it when onboarding becomes procedural instead of developmental. They notice when fewer people can explain why something works, not just how to run it.

They don't reject automation. They design around it.

I've seen this in plants where leaders intentionally invite people to question the setup, not just operate it. When someone asks, "Why is this done this way?" and the response is, "What do you think?" something shifts. Ownership returns. Engagement returns. And the improvements that follow tend to be practical, not theoretical.

One operation addressed this by introducing safety mentors, not to enforce compliance, but to create a place where uncertainty could be voiced without embarrassment. New hires weren't evaluated in those conversations. They were listened to. That design choice didn't make people braver. It made judgment usable.

As one plant manager explained to me, "Every time we add a new automated system, I ask what decision we're removing from people. Then I ask, 'Where are we giving them a different decision to make?' If we're not adding judgment somewhere else, we're setting ourselves up to fail when something breaks."

This path is harder to justify in pure efficiency terms. It introduces friction that metrics don't reward. But it produces a different kind of organization, one where people are still practiced at noticing, interpreting, and acting when the system doesn't know what to do.

And that difference doesn't show up on a dashboard. It shows up when conditions change and leadership finds out which kind of organization they actually built.

The Difference Isn't Visible on Dashboards

Both organizations hit their targets. Both show strong performance. Both report improvement. On paper, they're indistinguishable. If you're reading this and thinking, *We hit our numbers, but this still feels uncomfortably familiar*, that's not a coincidence. That's recognition. But when something unfamiliar happens, the difference is immediate.

In one organization, people wait for direction. They look to the system for guidance. They escalate uncertainty. They defer to authority. They need proof before they act.

In the other, people move. Not recklessly, but responsively. They trust their observations enough to interrupt momentum. They interpret signals before they become crises. They carry judgment that the system depends on but doesn't create. Not because they're more motivated. Because their roles still demand independent thinking.

At Oakridge, the path was set long before Line 4 failed. Every efficiency improvement that removed friction also removed a moment where judgment was required. Every optimization that reduced variance also reduced exposure to the conditions that build capability. By the time Denise wanted people to step forward, the organization had already trained them not to, without ever meaning to.

What Oakridge Mistook for Progress

Oakridge believed it was becoming more advanced. The numbers were strong. The efficiency gains were real. What leadership didn't see was that they were narrowing the space where judgment could develop. Each improvement reduced the need for interpretation. Each automation layer made compliance easier than thinking. Nothing about that felt wrong.

Until the morning when a bearing started to fail in a way the system couldn't recognize, and the only person who heard it had learned that hearing didn't count.

This Wasn't Locked In

Nothing about Oakridge's future was predetermined. Even three years into automation, after Frank retired, with training shortened and supervision reduced, the trajectory could have been altered. But that would have required leadership to ask a different question.

Not: How do we make the system better?

But: What is the system teaching our people?

Because that question shifts focus from performance to capability. From output to resilience. From what works today to what will work when today's assumptions no longer hold. Oakridge never asked that question. The system kept getting better. And the organization kept narrowing its margin for error.

Where This Leaves Us

Before we talk about rebuilding judgment, we need to be precise about what was lost. It's about whether your operation still produces people who can respond when the system doesn't know what to do. Because the most dangerous moment in a highly automated organization isn't failure. It's continuity.

When everything keeps working, and no one is learning anymore. That's when erosion happens fastest. That's when the gap between performance and capability widens the most. And that's when organizations drift, without realizing it, toward the moment when they will need judgment most and discover it's no longer there.

Stop reading. Right now. Go to your plant and spend thirty minutes watching where people defer to the system without questioning it. Don't intervene. Don't explain. Just watch.

If nobody defers, if people actively interpret, question, and pause when things feel off, congratulations. You are rare. If everyone defers? Welcome to the problem. Everything in the next section is about how you fix what you're about to see.

Part Two
The Smart Plant Framework™

By now you understand how judgment erodes. You've seen it happen at Oakridge. You've traced how reasonable decisions compound into fragility. You've recognized the patterns in your own operation.

What you need now is a framework for building something different.

The Smart Plant Framework™ is built on three core principles:

First, human judgment is infrastructure.

Not a soft skill. Not culture. Infrastructure. The same way you design, protect, measure, and maintain critical systems is how judgment has to be treated. Because when judgment fails, everything else eventually fails with it, no matter how advanced the technology looks.

Second, judgment doesn't survive on good intentions alone.

For people to use their judgment, the system must allow it. That means giving them the authority to act before certainty exists, the timing that makes early intervention legitimate, and the right to interrupt a process without asking permission. It also means protecting people when their judgment proves correct and building learning systems that don't require a major failure just to teach a lesson. Remove even one of those, and judgment retreats into silence.

Third, smart plants don't happen by accident.

They're deliberately designed. Leaders have to build operating architecture, protection mechanisms, and governance that support judgment under pressure before automation makes human capability optional.

These three principles form the foundation of everything that follows. The next seven chapters will show you how to operationalize them:

1. Designing work that builds judgment.
2. Protecting work when optimization pressures mount.
3. Scaling work across your operation without sacrificing the efficiency that keeps you competitive.

This isn't about choosing between technology and people. It's about designing systems where both amplify each other.

"Your technology strategy is impacting your HR strategy. That's the heart of the question to me."

— *Will Healy, III, Director of Product and Industry Marketing, Teradyne Robotics*

CHAPTER SEVEN: Judgment as Infrastructure

In highly automated operations, effort is rarely the constraint. People show up. They follow the process. They trust the system. When things go wrong, it's not because no one cared or tried hard enough. It's because the one thing the work still requires, human judgment, wasn't available when conditions changed.

That absence doesn't start on the floor. It starts upstream, in decisions about how work is designed, which people are allowed to think, and where discretion has been quietly engineered out.

Today, in Smart Plants™, performance depends less on effort and more on what people notice when something doesn't go according to plan. That kind of thinking can't be mandated. You can't spark it with slogans. You can't schedule it into a team huddle. It only shows up if the system still requires it.

So it's worth asking yourself a few uncomfortable questions. In your operation, where are people expected to notice, interpret, and decide? Where are they expected to wait for the system to tell them what to do? And when something doesn't fit the model, who is actually allowed to stop the line, ask why, or act without permission?

Modern leadership isn't just about setting goals or removing roadblocks. It's about deciding where judgment still lives. When the system decides by default. And what happens when human instinct and system output don't align.

Those decisions are being made constantly. In budget reviews. In process approvals. In role redesigns. In technology deployments. They rarely show up on a leadership scorecard. But they determine whether judgment is treated as infrastructure, something that's designed, protected, and reinforced, or whether it slowly erodes while performance metrics continue to look just fine.

Because the system will always do what it's designed to do. The question is whether your people are still allowed to think when the system doesn't know what to do next.

Why "People Strategy" Isn't Enough

Most organizations point to hiring initiatives, training programs, engagement surveys, and leadership development pipelines. But here's what they're missing: The problem isn't that people aren't engaged. The problem is that the work itself no longer requires them to think. You can't train your way out of a design problem.

Judgment isn't a personality trait. It isn't created by encouragement or unlocked by better messaging. It's not something you hire for and hope transfers across the organization. Judgment is built, or broken, by the design of the work itself.

If the work requires interpretation, people build that muscle. If the work rewards compliance, people learn to stop questioning. If success means monitoring rather than intervening, judgment atrophies through disuse. The problem isn't that people lack potential. The problem is that the work no longer asks them to practice the capability the organization will eventually need.

Leadership Happens Before the Line Runs

Culture doesn't begin with operations. It begins with design. Every time a process is approved, a role is scoped, a budget line is trimmed, or an automation project is greenlighted, culture is being shaped, even if it doesn't feel that way in the moment.

I've watched leaders make these decisions for years – good leaders, thoughtful leaders – without realizing they were also deciding who is expected to intervene before the system alerts, which roles interpret versus escalate, what kind of disagreement is welcomed and what kind systematically disappears, and where experience is required versus where compliance is sufficient. But they are cultural decisions with long-term consequences. They determine whether people believe their judgment still matters.

In many plants, no one ever announces, "Your instincts don't count anymore." They don't need to. Every signal tells that story: through incentives, approvals, procedures, and the slow reshaping of roles that no longer ask for experience.

Leadership doesn't remove judgment on purpose. It removes the need for it. And once the need is gone, the capability follows.

> **"Your customers are humans. AI is not human."**
>
> *— Katie Smith, Marketing & Customer Experience Leader, Wild Path Consulting*

Control Is Not Capacity

Automation gives leaders a powerful feeling: control. Processes run smoothly. Outcomes become predictable. Dashboards show normal. Variance drops. The operation feels disciplined, professional, and mature. But control isn't the same as capacity. Control helps you manage what you've seen before. Capacity helps you adapt when something new shows up.

Many high-performing plants have exceptional control and diminishing capacity. They execute known processes flawlessly. They respond to familiar problems quickly. But when conditions change, when a supplier fails unexpectedly, when material behaves differently, when equipment degrades in a way the model wasn't trained to recognize, the organization discovers it no longer has the margin it thought it had. The bench strength to respond when the system doesn't know what to do has been steadily traded away for the consistency to perform when it does.

Leadership didn't plan for that tradeoff. They weren't careless. They were solving for what they could see: reliability, predictability, cost control. The question they never faced: Does the work still create judgment capacity by design? Without asking that question, the capability simply stopped developing.

The Infrastructure Metaphor

This is where most leadership conversations about judgment break down. Leaders treat judgment as a cultural aspiration. Something to encourage. Something to value. Something to recognize when it appears.

But culture is optional.

When judgment lives in culture, it depends on individual courage. On someone being willing to speak up, slow things down, or challenge what the system says, even when that creates friction. In that environment, judgment survives only when people choose to exercise it despite the way the work is designed. And when judgment is optional, optimization wins every time.

This is where leaders need to get honest. Many say they want people to use judgment. But what really happens when someone does? Are they supported for noticing something early or questioned for disrupting the plan? Are they trusted when they challenge a recommendation or reminded that "the data looks fine"? Those responses teach people very quickly whether judgment is welcome or risky.

Judgment has to be treated as infrastructure.

Infrastructure is what the system assumes will be there. It's what work is designed around. It's what can't be removed without weakening the entire operation. Power is infrastructure. Connectivity is infrastructure. Safety systems are infrastructure. Financial controls are infrastructure. Judgment belongs in that category, not because people are special, but because no system, no matter how advanced, can anticipate every condition it will face.

What Makes Something Infrastructure

Infrastructure has specific characteristics that distinguish it from preferences, values, or aspirations:

Infrastructure is mandatory, not optional. The system cannot proceed without it. When infrastructure fails, work stops. When judgment is infrastructure, decisions cannot advance without it.

Infrastructure is designed for, not assumed. You don't hope power will be available. You design the facility around reliable power. You don't hope judgment will appear when needed. You design work so that judgment must be exercised repeatedly.

Infrastructure is maintained, not motivated. You don't inspire electrical systems to keep functioning. You maintain them. You don't motivate people to use judgment. You create conditions where judgment is practiced, protected, and reinforced.

Infrastructure has clear ownership. Someone is responsible for ensuring it exists and functions. When infrastructure fails, accountability is clear. When judgment fails, someone must own why the conditions for judgment were not maintained.

When judgment is treated as infrastructure, the leadership conversation changes completely. The question stops being: "How do we get people to think critically?" It becomes: "Where have we designed work that no longer requires thinking, and what are we doing to protect judgment where it still matters?"

Where Judgment Actually Lives

Judgment doesn't live everywhere equally. It concentrates in specific zones where its presence or absence determines outcomes.

Judgment lives at boundaries. Between shifts, context gets handed off imperfectly. Between automated and manual tasks, ownership blurs. Between departments, priorities collide. Between past data and present conditions that don't quite match. These areas are uncomfortable. They're messy. They resist standardization.

Over time, many organizations try to smooth those boundaries out. They automate handoffs. They clarify ownership. They simplify interfaces. They standardize responses. Each improvement removes friction. Each improvement also removes a space where judgment once operated. And when those spaces disappear, judgment doesn't relocate to a safer place. It dissipates.

Judgment lives in the gap between what systems see and what they miss.

Automated systems are excellent at monitoring what they've been taught to recognize. They check conditions, compare readings to thresholds, and respond when limits are breached.

Judgment operates earlier. It notices before the numbers shift. Before the alert fires. Before the system decides there's a problem. It shows up as unease. As pattern recognition that can't yet be formalized. As instinct built on exposure to conditions the model has never encountered. That noticing rarely comes with proof. And when organizations design work that values only certainty, those early signals have nowhere to land. They don't trigger action. They fade into silence.

Judgment lives where systems reach the edge of what they know.

For routine situations, encoding judgment into systems works well. Rules are documented. Logic trees are built. Models are trained on historical data. But when conditions change, when something novel appears, those systems encounter the boundary of their competence. That's where human judgment becomes essential.

Not judgment as stored knowledge, but judgment as active interpretation. The ability to recognize that current conditions don't match past patterns. The ability to act on ambiguity before it resolves into certainty. Judgment isn't something you extract and store in a database. It's something people practice. When systems replace judgment rather than support it, that practice stops. Not all at once. Gradually. And by the time the organization needs judgment to handle novelty, the people closest to the work have lost the capacity to generate it.

The Difference Between Knowing and Noticing

Systems are very good at knowing. They track thresholds, correlations, and probabilities. They recognize patterns they've seen before. They apply rules consistently. They don't forget. They don't get distracted. That's valuable. It's why automation works.

But *knowing* is not the same as *noticing.*

Noticing is human. It's built through exposure, repetition, and attention. Over time, people develop what I think of as a sensory library, a mental catalog formed by thousands of small interactions with the work. Sounds. Rhythms. Timing. Smells. Subtle changes in behavior or performance that don't register as data points but register as concern.

Humans notice when a vibration feels different. When a delay doesn't quite make sense. When a process technically meets specification but no longer feels stable. When a vendor checks every contractual box but raises unease that can't yet be quantified. That kind of noticing doesn't come from training manuals. It comes from lived experience and repeated sensory input.

If judgment is alive in an organization, ambiguity sparks curiosity. People slow down. They ask questions. They investigate. Uncertainty is treated as information, not noise. If judgment has eroded, ambiguity gets brushed aside. People wait for data to confirm what they're already sensing. They defer to the system's assessment. They escalate only when thresholds are crossed. By then, options have usually narrowed.

This is where leaders need to pause and reflect. How many inputs are your people getting into their sensory libraries? How often do they hear, see, feel, and troubleshoot the work themselves? And how often is the system doing that noticing for them?

Because systems don't lose judgment. They never had it. Organizations do, when people stop getting the experiences that build it. Surprises don't happen because the system failed. They happen because no one was positioned to notice what the system was never designed to see.

How Work Builds or Starves Judgment

People don't develop judgment just because they're capable or motivated. They develop it because their work requires them to interpret incomplete information, weigh tradeoffs, and decide without certainty.

If you've been in your role for a while, think back to when you were new. Not the first week, but the first few years. What did it take to really learn the job? The mistakes you made. The situations no one had a checklist for. The times you had to figure things out by watching, listening, trying something, and seeing what happened next. That's where your judgment came from.

Judgment grows through exposure to variation, through small problems that don't come with instructions, through moments where people have to reason through cause and effect instead of waiting for the next prompt. When the work stops demanding those things, judgment doesn't level off. It deteriorates.

As automation expands, many roles shift in subtle but consequential ways. Work that once required diagnosis now requires acknowledgment. Tracing causes gives way to following recommendations. Paying attention to how a system behaves under stress is replaced by trusting the system to flag what matters. On the surface, this looks like progress.

The job feels calmer. There are fewer decisions. Errors appear to drop. Training time shortens. Productivity accelerates. But the thinking doesn't disappear. It gets deferred. Instead of being spread across dozens of small decisions made early, judgment concentrates into a handful of high-stakes moments later, when the system encounters something it can't resolve on its own.

By then, the people closest to the work are out of practice.

They haven't been exercising the muscle the organization suddenly needs them to use. The work hasn't required judgment until the moment when everything depends on having it ready. And that's the risk leaders rarely see coming. Organizations don't discover this gap during routine operations. They discover it during a crisis, when the stakes are highest and there's no time to rebuild what was slowly dismantled in the name of efficiency.

That's when leaders realize they've optimized their way into a corner that they can't think their way out of.

Why Training Can't Replace the Job

When leaders sense that judgment is fading, their first response is almost always more training. New courses. Refresher modules. Simulations designed to re-create edge cases. Leadership programs focused on critical thinking. That instinct makes sense. Training feels actionable. It signals investment. And when it's done well, it can be helpful.

But training only works when the job itself still requires judgment.

Judgment doesn't stick when it's learned in isolation. It develops through repeated exposure to real situations where interpretation matters and decisions carry consequences. When the work no longer asks people to interpret, weigh tradeoffs, or decide without certainty, training becomes theoretical. People return to the floor with new language and no place to use it.

The issue isn't that training fails. It's that training is being asked to solve a work design problem. You can't train people to think critically if their job rewards compliance. You can't develop interpretive skills if success means confirming what the system already decided. And you can't expect judgment to show up under pressure if the work never requires it during normal operations.

Training treats judgment as a knowledge gap. Work design determines whether judgment is needed at all. When those two aren't aligned, training doesn't reinforce judgment. It competes with the system. And the system always wins.

The Hidden Cost of Efficiency

Many of the decisions that starve judgment are framed as efficiency wins. Onboarding gets shorter because the system handles complexity. Manual checks disappear because sensors are more reliable. Roles are simplified so fewer people can run the same process. Supervision is reduced because AI handles real-time problem-solving. Authority narrows because centralized systems make better decisions.

Each choice makes sense on its own. Together they reshape what people learn.

New employees become fluent in execution but unfamiliar with cause and effect. They know how to respond but not how to anticipate. They learn which button to press, not what's happening beneath the interface. What quietly disappears in that process is context.

The moments where judgment transfers are the moments that efficiency trims first. Explaining why something matters, not just how to do it. Inviting someone into problem-solving instead of shielding them from it. Answering questions with history and reasoning instead of procedure. Over time, those moments are engineered out of the work. This is also where organizations lose something they rarely name until it's gone: the *why* behind the operation.

Founders and long-tenured employees carry more than technical knowledge. They carry the reasons certain decisions were made. Why this customer matters. Why that shortcut isn't worth taking. Why quality is non-negotiable even when the numbers say you could get away with less. That perspective doesn't live in SOPs or dashboards. It's passed down through stories, explanations, and shared experience.

When experienced employees leave, the organization doesn't just lose headcount. It loses the perspective it has no mechanism to preserve. Performance metrics don't flag this as a risk. Efficiency still looks like success. Output stays strong. Costs decline. Quality holds.

The erosion remains invisible because what's being lost isn't performance. It's capacity.

Reducing Error Versus Reducing Learning

Many systems are designed to minimize human error. That's necessary. Error is costly. Consistency matters. Reliability is valuable. But there's a difference between reducing error and reducing learning.

When work is designed so people can't make small mistakes, they also lose the chance to practice recovery. They don't learn how systems behave under strain. They don't build confidence in their ability to intervene. They don't develop the tacit knowledge that comes from navigating ambiguity. Over time, dependence replaces capability.

"The dirty, dangerous, demanding work goes to automation. Let them have it. We evolve to be cognitive thinkers."

— Brent Kedzierski, Manufacturing Technology Leader, kNot Today

People become excellent at operating within the system's boundaries. They perform well when conditions are stable. They follow procedures flawlessly. But when something unexpected happens, when the system encounters a condition it wasn't designed for, people hesitate. Not because they lack intelligence, but because they haven't practiced thinking outside the structure the system provides. That dependence feels safe, right up until the system encounters something it's never seen before.

The Trade Leaders Rarely See

What's being traded away isn't engagement or effort. It's exposure. Exposure to ambiguity. Exposure to variation. Exposure to situations where judgment has to be exercised before certainty exists. Those moments are uncomfortable. They slow things down. They introduce friction. They create variance. They resist standardization. They also build capability.

When leaders remove friction without considering what it teaches, they unintentionally starve the organization of the conditions that produce judgment. Friction is where learning happens. Variation is where adaptability grows. Ambiguity is where judgment develops. Smooth operations don't create those conditions. And when operations become too smooth, when every decision is algorithmic, every response is scripted, every outcome is predictable, the organization stops producing people who can handle surprises.

Where Judgment Must Be Protected

Not all work needs to preserve judgment equally. Some roles should be procedural. Some tasks should be automated. Some decisions should be scripted. But certain spaces must remain interpretive by design.

Before commitment: Once resources are locked. Schedules are set. Materials are consumed. Production is released. Options narrow rapidly. Judgment placed before commitment prevents cost. Judgment placed after commitment explains it.

Where signals conflict: When data disagrees. When departments see the situation differently. When system recommendations don't align with operational reality. These are moments that require interpretation, not calculation. If no one is explicitly expected to make sense of conflict, the system resolves it by default, often incorrectly.

Where consequences compound: Some local decisions create non-local effects. A scheduling choice that optimizes today creates fragility tomorrow. A maintenance deferral that saves budget this quarter creates risk next year. A quality decision that meets specifications creates downstream customer dissatisfaction.

These decisions require judgment that goes beyond immediate efficiency. If judgment doesn't live where those tradeoffs are made, the organization optimizes locally and fails systemically.

At boundaries: Shift handoffs. Department interfaces. Automation-to-manual transitions. These are the spaces where context is lost, ownership blurs, and problems fall through gaps. If judgment doesn't live at boundaries, surprises emerge from places no single system was designed to monitor.

This Is the Leadership Work Now

The job has changed. Leadership today isn't just about driving execution or boosting morale. It's about building environments where judgment can survive optimization. Where noticing matters more than reacting. Where interpretation is expected, not simply optional. Where disagreement with a system isn't resistance, but a feature. Where experience lives in the work, not just in documentation.

This kind of leadership doesn't show up during a crisis. It shows up long before, when roles are defined, systems are scoped, budgets are approved, and decisions are made without witnesses. Leadership shows up in what the system asks of people, and what it no longer does.

What This Changes

If leadership is a design function, then familiar questions start to shift. Questions about culture become questions about what the system actually signals matters. Questions about resilience become questions about where judgment is still being built, and where it has quietly stopped developing.

This isn't an argument against automation. It's a call to lead it with precision.

When judgment disappears from the design, wishing doesn't bring it back. Simply noticing its absence doesn't restore it. Judgment has to be built into the system deliberately, practiced through the work, and protected over time. Not as a value statement. As infrastructure.

Because surprise is inevitable. Systems will encounter conditions they weren't designed for. Reality will eventually outpace what automation can resolve on its own. When that happens, judgment isn't optional. It's load-bearing. Treat it that way. Or accept that when the system finally needs it, the capability won't be there.

So, take a look at your last three unplanned stoppages. For each one, ask yourself: *Did a person catch this or did the system?*

If the system caught all three, the question isn't whether your plant can run without judgment.

It's whether your people still know how to use it.

And here's the more difficult follow-up: When was the last time someone in your operation acted on instinct before the data confirmed they should? When was the last time someone slowed the line because something felt wrong, even though every indicator was green?

If you can't remember, judgment isn't missing. You've designed it out. And the work in Part Two is about designing it back in before you need it and then discovering it's gone.

"So much of what they do for those thirty years is just intuitive. If you ask them, "How do you do this?" they say, "I don't know, I just do it."

— *Lisa Sanderson, Vice President Marketing and Strategy, Gleicher Manufacturing Corp*

CHAPTER EIGHT:

Reading the Erosion

You know judgment is eroding. The question is: where? Walk your floor tomorrow and watch what happens when someone notices something the system doesn't confirm. Don't explain it. Don't intervene. Just watch. That's your diagnostic.

Most leaders don't realize judgment has faded until something unexpected happens: a failure the model didn't predict, a response that feels slow when it should be instant, a team that hesitates when action was obvious

But that's not the real surprise. The real surprise is why it wasn't seen. More precisely, what had gradually stopped being looked for.

The Myth of Sudden Loss

Judgment doesn't disappear in a dramatic moment. It wanes while leaders are being praised for running a tight operation. When I talk to plant managers about this phase, here's what they tell me: "We're finally getting ahead of things." Fewer fire drills. Less pushback. People come up to speed faster. The floor runs better. Everyone treats this like a win.

The fade happens incrementally – not through neglect, but through reasonable choices that make the work run smoother. By the time leadership notices judgment is missing, the conditions that once required it are already gone.

Which means the question isn't whether your organization values judgment. The question is whether your operation still creates the conditions that require it.

Many operations don't. And the more successful an operation appears, the harder it becomes to see.

That's a different kind of diagnostic. It requires looking not at outcomes, but at behaviors. Not at what people deliver, but at what the work makes unavoidable. Not at the signals that confirm performance, but at the small, easily overlooked ones that reveal whether judgment is being practiced – or noiselessly optimized away.

Where Leaders Usually Look First

When something feels off, most leaders turn to familiar indicators: performance metrics, compliance reports, engagement scores, training completion rates, turnover statistics. These tools are useful. They're also where leaders have been trained to look for years, even when the real issue lives somewhere else.

They tell you whether processes are being followed, systems are stable, and expectations are being met. What they don't tell you is whether judgment is still active. Judgment doesn't leave an audit trail. It shows up in behaviors that rarely trigger alarms and are often misread when they do.

> **"People want to be heard, appreciated, and know their employers care enough to invest in them."**
>
> *— Darrin Mitchell, Chief Marketing Officer, Manufacturing Masters*

Early Signals That Rarely Raise Concern

When judgment erodes, the earliest signals rarely look like risk. They look like improvement.

Escalations decline, even when conditions feel ambiguous. Fewer interruptions reach leadership. Fewer concerns are raised without complete data. Issues surface later in their lifecycle, packaged as certainties rather than questions.

Disagreements with system recommendations taper away. Overrides remain technically available, but they occur less often over time. Leaders often interpret this as growing confidence in the system. What's harder to see is when people stop challenging recommendations altogether, even when something feels off.

Action occurs only after alerts, never before. In environments where judgment is still active, people intervene ahead of thresholds. They slow work when early drift appears, even if performance metrics remain within bounds. As judgment erodes, intervention becomes reactive. People wait for confirmation before acting.

New hires become productive quickly, but struggle when conditions deviate. The unease sounds remarkably consistent: "They're great, until something's not in the playbook." New employees hit targets fast and follow procedures well. When the system doesn't tell them what to do, they stall – not from lack of effort, but because the work never required them to practice deciding.

Experienced employees are consulted less and speak up less. As systems become authoritative, experience loses standing. Observations that once carried weight are deprioritized if dashboards don't confirm them. Over time, seasoned operators conserve their insight. When experience goes quiet, the organization loses its early warning system.

> **"The only way to learn is in the shop, working on the machines. There's going to be fewer people around to train because everybody's retiring."**
>
> — *Noah Graff, Manufacturing Journalist, Today's Machining World*

Cross-functional problem-solving declines. A maintenance tech notices a subtle change in machine behavior. Production sees output still in spec. Engineering never hears about it because there's nothing formal to escalate. Everyone does exactly what their role requires. The signal never crosses the room.

Judgment often operates at the intersections: between production and quality, maintenance and engineering, supply chain and operations. As roles become more scripted and work more compartmentalized, these informal exchanges happen less often. Departments improve locally. Signals don't travel until problems are already formalized. By then, they are harder to solve, more expensive to fix, and often visible to customers before leadership knows they exist.

Everything reads as improvement. Which is exactly why erosion continues unchallenged.

The Question That Changes What You See

When something unusual happens – a quality deviation, a process anomaly, a machine behaving differently – ask yourself: *Who notices first?* Is it a sensor or a person? And when a person notices early, before the data confirms it, what happens next?

Are they encouraged to pause the process? Are they given space to explore the concern without proof? Are they protected if their instinct turns out to be wrong?

If people consistently wait for confirmation before acting, judgment is already slipping. Unease doesn't arrive with data. Judgment operates in the space before certainty exists. If that space has been closed, if uncertainty is treated as noise rather than signal, then judgment has nowhere to operate. And once people learn that acting on incomplete information carries more risk than waiting for the system to confirm the problem, they stop using the very capability the organization will eventually need most.

Where Judgment Disappears First

Judgment rarely fades evenly across an organization. It disappears first in the places that feel the most polished.

Highly automated lines are often the earliest sites of erosion. The more sophisticated the automation, the less human interpretation is required for normal operation. That is the design intent. That is why automation delivers value. But the side effect is subtle and cumulative: people spend more time monitoring and confirming than interpreting and intervening. Over time, the work stops requiring judgment to proceed, and a capability that isn't exercised begins to fade.

When something appears outside the model's training, the capability to respond is often still present, but the confidence to use it is not. You can see this play out in real time. There's a pause before someone overrides a recommendation. A glance toward a supervisor for permission to trust their own read of the situation. A sense of relief when the system finally catches up and confirms what they already suspected. These are not signs of incompetence. They are signs that the work itself has trained people to defer rather than decide.

Simplified roles show a similar pattern. Simplification is usually framed as a win: fewer decisions, clearer responsibilities, easier staffing, faster training. And in many respects, it is. But simplification also removes complexity, and complexity is where judgment develops. When a role is reduced to following a sequence, success no longer requires interpretation. The role produces execution, not discretion. Over time, people become very good at doing what is expected and less practiced at deciding what to do when expectations no longer fit.

Processes that run with minimal manual interaction carry the same risk. Smooth processes are celebrated for good reason. Low variance. High consistency. Minimal intervention. But smoothness also means fewer opportunities to practice judgment. When a process runs perfectly for months, people forget how to troubleshoot it. They forget what it sounds like under stress. They forget the early signals that precede failure because the work no longer requires them to notice.

When that process eventually encounters something novel, the people responsible for it may struggle to recognize what matters. The process you are most proud of can quietly become the one least prepared for surprise.

The Assumption That Creates Risk

Many leaders hold an unspoken belief: *If something truly unexpected happens, our people will rise to the occasion.* Sometimes they do. But that belief only holds when people have been practicing for that moment all along, when their roles still require them to observe, interpret, question, and act before being told.

If the work no longer asks that of them, the issue is rarely ignored outright. More often, it is deferred. And sometimes, that deferral is the failure. The window for effective intervention is often narrow. By the time the system confirms there is a problem, by the time leadership engages, by the time the data is clear, the organization is already managing consequences instead of preventing them.

The Most Revealing Test

Here's a scenario that reveals more than any engagement survey or performance review: tomorrow, a system recommendation conflicts with what an experienced operator feels in their gut. The data says proceed. The operator says hold up. Which one wins?

You already know the answer.

If the default is "trust the system unless the operator can prove otherwise," then judgment has already been sidelined. Judgment doesn't arrive with proof. It arrives as uncertainty. As discomfort. As a sense that something isn't right before there's language for why.

A plant manager I worked with described it this way: "You can feel the temperature in the room. When it goes flat – not tense, not chaotic, just flat – that's when I know people have stopped speaking up."

Leaders often misread that flatness as control. The floor looks orderly. The operation feels settled. There's no drama, no escalation, nothing demanding attention. What's happening is more concerning. People have learned that unease without data doesn't count. That observations without confirmation aren't welcome. That the safest move is to wait.

And once judgment needs proof before anyone will take it seriously, judgment will always arrive too late.

The Diagnostic Most Leaders Avoid

There's a simple way to tell whether judgment is slipping: track how often people disagree with automated recommendations. Not how often they're right. How often they push back at all.

A healthy operation maintains selective disagreement. People rely on the system most of the time, but they are practiced enough to recognize when context matters more than the model.

When disagreement drops toward zero, one of two things is happening. Either your AI has achieved perfection (it hasn't), or your people have stopped believing their judgment carries standing. The second is far more common. And far more dangerous.

What Frontline Silence Actually Means

When experienced operators stop raising concerns, leaders often read it as alignment. "They trust the system." "The operation is running as planned." Sometimes that's true.

More often, silence means people have learned that unease without data creates friction with no effect. So they conserve it. That shift, from practiced judgment to passive monitoring, is rarely tracked and almost always misread. And by the time it shows up in outcomes, the authority has already moved.

The Gap Between Competence and Readiness

Performance metrics measure competence: Can people do the job as designed? What they don't measure is readiness: Can people respond when the job encounters conditions it wasn't designed for? Competence is visible. It shows up in productivity, quality, and efficiency. Readiness stays invisible until the moment you need it.

You can run a highly competent operation with very little readiness. People execute tasks cleanly, follow procedures precisely, and hit every target. Then something genuinely unfamiliar shows up and the response slows. Not because people lack skill, but because the work never required them to practice deciding under uncertainty.

Readiness is built through exposure to ambiguity, variation, and incomplete information. It develops when people are expected to make calls before outcomes prove whether those calls were correct. When work is designed to remove those conditions, it also removes the conditions that build readiness.

This is usually where leaders push back.

"So you're saying we should slow down?"

"So automation is the problem?"

No. What I'm saying is simpler and more difficult: judgment doesn't survive by accident. If you want people ready when systems fall short, the work has to keep them ready. Readiness isn't something you stockpile during smooth operations and access later. It's a muscle. If the work doesn't use it, you lose it.

What This Leaves You With

You don't need a major failure to know whether judgment is eroding. You need to pay attention to how your operation handles ambiguity. When something feels off, do people slow things down or do they wait for confirmation? Do concerns surface before data makes them defensible or only after the system gives permission? When experience conflicts with an algorithm, which one carries weight? And when someone disagrees with a system recommendation, is that treated as insight or as inconvenience?

If those behaviors are becoming rare, judgment hasn't disappeared. You've stopped exercising it.

Here's the part most leaders miss: when a situation finally demands judgment, recognition won't bring it back. You can't summon readiness in the moment it's needed. It shows up only if the work has been building it all along, day after day, long before anything looks urgent.

Judgment isn't a crisis response. It's an everyday byproduct of how work is designed. And when the work stops requiring it, operations often become more efficient and more fragile at the same time. Every indicator still looks strong – right up until it doesn't.

So this is the choice in front of you. You can keep improving what's easy to measure and hope you never face what isn't. Or you can start designing work that builds the capability you'll eventually need.

The next chapters will show you how. But only if you're willing to look honestly at what smoother operations might be removing.

"How can we use automation to eliminate the physical demands but keep the mental parts? People have so much knowledge."

— *Will Healy, III, Director of Product and Industry Marketing, Teradyne Robotics*

CHAPTER NINE:

Designing for Judgment

If Chapter Eight left you with an uneasy feeling, that's the point. Once you can see how judgment fades in day-to-day work, the next question isn't *whether* it's happening. It's why it keeps happening even in well-run, well-intentioned organizations.

The answer isn't culture. It isn't effort. And it isn't leadership commitment in the abstract. Judgment erodes because of how work is designed.

Not in one big decision and not in a moment anyone flags as risky. It happens through the accumulation of reasonable choices: an approval removed to speed things up, a role simplified to reduce variability, a system given authority because it performs consistently, a process changed so fewer people have to think about it. Each decision makes sense on its own. Together, they determine how – and whether – judgment shows up when something doesn't go as planned.

Design is where judgment lives or dies, and it happens long before you notice. Every workflow you approve, every role you simplify, every automation project you greenlight – you're making a choice about who is expected to think and who is expected to comply. The work itself trains people faster than any culture deck ever will. If the work rewards waiting for certainty, your people will wait. If it rewards moving without interpretation, they'll move. And by the time you need them to use judgment, they'll have spent months or years learning that doing so is risky. You don't lose judgment in a meeting. You lose it in a thousand small design decisions that nobody flagged as dangerous.

So the question going forward isn't whether you value judgment. It's what kind of judgment your operation is being designed to produce and under what conditions it will still be allowed to act.

That's what this chapter is about.

Design Happens With or Without You

Most leaders do not think of themselves as designers. They think of themselves as operators, problem solvers, people responsible for keeping things running. Design sounds intentional and creative. What actually shapes judgment feels far more ordinary.

Judgment is never shaped by what leaders say they value. It is shaped by what the work makes unavoidable.

If a role requires interpretation, people learn to interpret. If it rewards compliance, people learn to comply. If the system decides first, people learn that their deciding is unnecessary. No announcement is required. No strategy deck. The work teaches people exactly how much – or little – thinking is expected of them.

This is the part that is easy to miss. You do not have to decide to remove judgment for it to disappear. Allowance is enough. Approval is enough. Letting a design change pass without asking what it removes from human discretion is enough.

Every process tweak, every role simplification, every automation decision teaches the organization how much judgment is expected and where it no longer belongs. You may not be dismantling judgment intentionally, but if you are not actively protecting it, the work will redesign itself toward efficiency. Efficiency, left alone, always chooses the path that requires less human interpretation. Less interpretation means less practice. Less practice means less capability when conditions stop cooperating.

The Silent Power of Defaults

In most organizations, judgment is shaped less by policy than by default.

Who is expected to escalate and when. Whether system recommendations are treated as guidance or instruction. How much uncertainty is tolerated before someone is told to move on. What happens when a person slows down the process. These do not feel like leadership decisions. They feel like "how things work around here." That is exactly why they are powerful.

I was reminded of this during a safety keynote I sat through at an industry conference. The speaker talked about the difference between the written safety policies and what actually happened on the floor. The policies were clear and well-intentioned. But what governed behavior was the norm. What people really did. What was tolerated. What was expected. The rulebook said one thing. The work taught another. And everyone in the room knew which one mattered more.

Judgment works the same way.

Defaults teach faster than any training program ever will. They teach when it is safe to speak and when it is smarter to stay silent. They teach which kinds of thinking are welcomed and which carry risk. They teach whether disagreement with the system is insight or inconvenience and whether noticing something the data does not yet confirm is valued or dismissed.

Over time, those defaults harden into culture. Not the culture described in values statements, but the one people experience when no one is watching.

Here is what makes the pattern so difficult to see. You probably did not choose most of these defaults. They emerged. Someone approved a process. Someone optimized a workflow. Someone removed a step that "didn't add value." No one set out to erode judgment. But small choices compound. Eventually, the defaults are running the operation, whether you realize it or not.

This is a useful pause point. Not to revisit your values, but to look at the work itself.

Where does the work require people to interpret before the system confirms? Where does it allow someone to slow things down without penalty? Where does judgment still have standing, and where has it already been designed out?

Whether you claim the role or not, you are already designing for judgment. The only question is what your design is teaching people to do when something does not fit the model.

What Gets Rewarded Is What Sticks

Judgment does not disappear because people forget how to think. It recedes because systems make clear which kinds of thinking are worth the effort. When success means staying on schedule, people keep things moving. When praise follows uninterrupted flow, people avoid raising issues. When system output is treated as final, human insight becomes optional. Not forbidden, just unnecessary. And what is unnecessary does not stay sharp.

> **"You can't just plop down a new process, new technology, and change everybody's working life. People might be doing it for 15-20 years."**
>
> — *Vinny Maurici, COO, Syndigo*

Leaders rarely discourage judgment directly. Instead, they reward outcomes that do not require it. Perfect uptime. Zero deviations. Clean handoffs. All of those can be achieved through compliance alone. When compliance is consistently reinforced and interpretation introduces friction, people do not need a memo explaining the shift. They adjust.

People do exactly what the system asks of them. If it asks for monitoring, they monitor. If it asks for escalation, they escalate. If it asks for confirmation, they confirm. The issue is not that people follow the system. The issue is that following has become the only behavior the system consistently recognizes.

Here is the part leaders often underestimate. Your most experienced people notice this immediately. The operators with the best instincts and the longest memory are paying attention to what gets acknowledged and what disappears without comment. Over time, they stop offering what the system does not seem to want. Not out of frustration or disengagement, but out of self-preservation. Offering judgment that goes unrecognized or complicates the flow is exhausting. So they reserve it. And the organization slowly loses access to exactly the insight it will need when conditions stop cooperating.

The Trade No One Puts on the Agenda

Every operational decision makes a trade. The risk is not the trade itself. The risk is failing to name it.

During a safety tour at a fastener manufacturer, leadership debated whether to allow earbuds on the floor. On the surface, the discussion focused on hearing protection, worker preference, and retention. Hiring was tight. Enforcement would create friction. The reasons to say yes were immediate and tangible. The reasons to say no were distant and abstract, including the cumulative risk that earbuds used in high-noise environments create gradual hearing loss that proper PPE is designed to prevent.

What made this a judgment design issue was not the decision itself, but what followed. Once the exception was approved, who owned monitoring whether it still made sense? Who was expected to revisit the decision as conditions changed? Who was responsible for noticing when the tradeoff no longer held?

No one.

The approval removed discomfort for supervisors who would have had to enforce the rule. It also removed the ongoing practice of weighing risk, revisiting assumptions, and asking whether yesterday's compromise still made sense today. Once exceptions stop feeling like exceptions, they stop being questioned. That is not a policy failure. It is a judgment design failure.

Speed trades away exploration. Control trades away surplus capacity. Efficiency trades away margin for error.

These trades are not mistakes. They are often necessary. The problem arises when they are made implicitly and never revisited. Over time, organizations build systems that perform beautifully under known conditions and teams that look polished but have not practiced deciding when conditions change.

Notice what rarely happens. Executive meetings do not ask what capability is being traded away for this efficiency gain. Budget reviews do not ask what judgment is being removed to achieve this cost reduction. Technology deployments do not ask where interpretation is being preserved as automation expands.

When those questions are not asked, efficiency wins by default. And judgment slowly gives way – not because anyone chose to remove it, but because no one chose to protect it.

Where Judgment Must Live

If judgment has been engineered out, the next question isn't how to restore it everywhere. It's where judgment belongs. Preserving judgment in the wrong places is just as ineffective as losing it entirely. Judgment doesn't need to exist everywhere. Systems should handle the repeatable, the routine, the predictable.

At Commitment Points

Judgment matters most before commitment. Once resources are locked, schedules set, production released, or maintenance deferred, options narrow rapidly. Reversing course becomes expensive. Judgment placed before commitment prevents cost. Judgment placed after commitment explains it.

Most organizations allow commitment to happen automatically. Systems sequence work, algorithms optimize schedules, and production releases according to plan. If no human is required to interpret conditions before commitment, judgment arrives too late. This is why failures feel surprising even when early signals existed; the signals appeared before commitment, but the response came after.

Ask yourself: In your operation, who has the authority to pause commitment in the face of incomplete information? If the answer is "nobody" or "only in extreme cases," you've designed judgment out of the moment it matters most.

At Boundaries

Boundaries are where judgment earns its keep – between shifts, between automated and manual tasks, between departments, between what the system expects and what actually shows up. Most breakdowns don't begin inside processes. They begin between them. Systems excel within their lane. Real problems show up where lanes meet.

Because boundaries are messy, organizations try to clean them up. They standardize handoffs, automate transitions, and streamline interfaces. Each improvement removes friction. Each improvement also removes a space where judgment once operated. When boundaries become too neat, early warning signals disappear.

Think about your most recent operational surprise. Where did it originate? Chances are, it showed up at a handoff, a shift change, or an interface between systems. Someone probably noticed something off but didn't have a clear path to escalate across the boundary. The cleaner your boundaries look on paper, the more you should worry about what's being lost in translation.

Where Signals Conflict

Judgment is essential when inputs disagree – when data says one thing and experience suggests another, when system recommendations don't match operational reality, when metrics are green but something still feels wrong. These moments require interpretation, not calculation. If conflict is resolved by default, usually by deferring to the algorithm or authority, judgment has nowhere to operate.

Conflict isn't noise. Conflict is information. When judgment doesn't live where signals conflict, organizations optimize for coherence rather than accuracy. They choose the answer that makes the dashboard look clean over the answer that might be true.

Here's the test: When was the last time someone in your operation successfully challenged a system recommendation based on gut feel? Not *after* the system failed but *before* it failed. If you can't think of an example, that doesn't mean your systems are flawless. It's because your people have learned not to bother.

Where Consequences Compound

Some local decisions create non-local effects. Scheduling optimization boosts today's throughput but creates tomorrow's vulnerability. A maintenance deferral saves budget now and introduces risk later. A quality decision meets specs but erodes customer trust downstream. If judgment doesn't live where these tradeoffs occur, organizations optimize locally and fail systemically.

Every subsystem performs. Every metric looks normal. And the interaction between subsystems creates vulnerabilities that no dashboard reveals. You won't see this coming in your KPIs. You'll see it when a customer calls, when a line goes down unexpectedly, or when three small problems arrive simultaneously and your team has no practice managing the collision.

The hard question: Does anyone in your organization have the mandate to slow down a local optimization because they are worried about system-level consequences? Or does every department improve independently, hoping someone else is watching the whole?

Why Clean Isn't Always Safe

When work gets cleaner, interruptions disappear. Fewer disagreements, fewer pauses. That can look like progress. But judgment thrives in friction, not chaos, not dysfunction, just enough resistance to require interpretation. When friction disappears, judgment stops being practiced.

Smooth operations handle the known exceptionally well. They fail when something unfamiliar appears. And here's what I need you to sit with for a moment: If your operation feels smoother than it did five years ago, that's not automatically a win. It might mean you've optimized away the very capability you'll need when smooth stops working.

Judgment Is Contextual, Not Universal

Not every role requires the same level of judgment. Some roles should execute. Some should optimize. Some should move fast. But certain roles must remain interpretive by design – roles where signals conflict, where timing matters more than certainty, where consequences extend beyond the immediate decision, where context changes faster than procedures can keep up.

Strip down those roles to monitoring and escalation, and the operation may still function. It just won't be resilient. Think about your critical roles right now. Are they designed to interpret or execute? If you're not sure, that's your answer.

What Preserving Judgment Actually Means

Preserving judgment isn't about resisting automation. It's about deciding where the system is NOT allowed to proceed without a human interpreting what's actually happening. Not monitoring. Not confirming. Interpreting. If you can't name three places in your operation where the system must stop and wait for human judgment, you've already automated it away.

These roles look inefficient on a spreadsheet. They introduce variance, create friction, and slow what could otherwise move faster. They are also the roles leaders defend after something goes wrong, when the question shifts from *why did this happen* to *why didn't anyone speak up sooner*. Those roles preserve the organization's ability to notice what the system doesn't.

So here are the uncomfortable questions: Do you still have roles like this? Have you protected them deliberately? Or have you optimized them all into compliance?

The Tradeoff Leaders Must Name

Every organization lives inside tradeoffs: speed versus sensemaking, consistency versus adaptability, control versus judgment, efficiency versus resilience. You can't maximize all of them. Most leaders pursue what's measurable until judgment is what gets squeezed out. Then they're surprised when no one acts during a crisis.

The surprise isn't that people didn't act. The surprise is that the system never trained them to.

> **"Think about the person who moved across the world. That person has grit you can't teach."**
>
> — *Dave MacDonald, Manufacturing Executive, Better Together Group*

Naming the tradeoff changes the conversation. Instead of asking, "How do we get people to think more critically?" you start asking, "What are we willing to sacrifice to preserve judgment where it matters?" That's a harder question. It requires admitting that you can't have perfect efficiency and perfect resilience. It requires choosing what you're willing to give up. But it's the only question that leads to deliberate design instead of default drift.

And if you're not willing to make that choice explicitly, the operation will optimize itself. It won't ask permission, and it won't tell you what it traded away.

What Protection Actually Requires

Protecting judgment isn't about resisting automation or giving people free rein. It means making three specific design commitments and defending them against constant pressure to optimize:

First, the authority to interrupt. People must be able to pause momentum instinctively, not only when the data supports it. If someone needs three levels of approval to stop a line because "something feels off," you've designed judgment out of the moment it matters most.

Second, roles explicitly designed to interpret. Some positions must exist to make sense of ambiguity, conflict, and edge cases, not to execute procedures or monitor dashboards. These roles look inefficient on paper because they introduce variance and create friction. That's the point. They preserve your organization's ability to notice what the system doesn't.

Third, friction that forces thought. Enough resistance in key decision points to ensure interpretation happens before commitment. This isn't bureaucracy for its own sake. It's the difference between catching problems early and explaining them later.

Here's what makes this so difficult: every quarter, someone will propose streamlining one of these protections. The case will be compelling. The efficiency gains will be real. And unless you can articulate why that friction exists, you'll approve the change. Then another. Then another. Until the day you need judgment and discover it's been optimized away, one reasonable decision at a time.

The Line You Can't Cross Twice

Judgment usually disappears where leaders aren't looking. In handoffs, in edge cases, in roles labeled "low-risk," and in decisions that feel tactical rather than strategic. Once judgment is removed from those spaces, bringing it back requires more than good intentions.

I've worked with leaders who realized too late that they had traded away their bench strength. They tried training programs, engagement initiatives, and empowerment campaigns. None of it worked. The work itself had not changed. The system still rewarded compliance over interpretation. The pressure to maintain the schedule still outweighed the permission to pause. You can't train your way out of a design problem.

Here's the part that's uncomfortable. Most of those leaders could point to the exact decisions that made sense at the time. The approval that kept things moving. The simplification that reduced friction. The exception that felt reasonable. None of it looked reckless. It just crossed a line they didn't know they couldn't cross twice.

What This Leaves You With

The most important leadership question isn't whether judgment matters. Everyone agrees it does. The question is whether your operation is still building it. Not whether people are capable of thinking, but whether the work requires them to. Not whether you value interpretation, but whether success depends on it.

Walk your floor tomorrow and watch how decisions get made. Not during a crisis, but on an ordinary Tuesday. Who is interpreting and who is executing? Who has permission to slow things down, and who is being measured on throughput? Where does the system expect people to think, and where does it only expect them to confirm?

Those observations will tell you more than any engagement survey ever could. Because if judgment isn't being practiced in everyday work, it won't be available when conditions change. And the next time your operation faces something the system wasn't designed for (and it will), you'll find out whether you've been building capability or just managing compliance.

Here's your homework: Pick one process that runs smoothly. One that you're proud of. One with clean metrics and rare exceptions. Now ask yourself: If the system failed right now (not alerted, just failed), who would notice first? And how long would it take? If your answer is measured in hours instead of minutes, that smooth process is more fragile than you think.

"As much as 97% of lean efforts failed to achieve what they set up to accomplish. Most start their lean efforts with the wrong purpose; they focus on the metrics as opposed to focusing on the holistic approach."

— *Scott Gauvin, Founder/CEO, Macresco*

CHAPTER TEN:
Why Retrofit Fails

When leaders finally realize judgment is thinning, they do what they've been trained to do: they launch a program. Another training module. Another town hall. Another poster about speaking up. It feels decisive because you can put it on a slide. It feels responsive because people in the room nod. And it feels like leadership because you spent money. Here's what you need to hear: it's theater. Expensive, well-intentioned theater that makes you feel better while the actual problem gets worse. You cannot train your way out of a system that's been designed to punish thinking. And every dollar you spend pretending you can is a dollar you're not spending on the redesign that might actually work.

Not because the programs are poorly designed. Not because people don't care. Not because the message isn't sincere. Judgment doesn't fade from lack of awareness, and no program can restore what the work itself no longer requires.

The False Comfort of Programs

Programs feel productive. They're visible, trackable, and easy to announce. They give leaders something to point to as proof that the issue is being addressed: leadership development modules on critical thinking, communication campaigns emphasizing empowerment, training sessions on decision-making under uncertainty, and town halls where executives encourage people to speak up.

These initiatives aren't wrong. In some contexts, they help. I don't say this lightly. Training is part of how I make my living, and I believe deeply in its value. Yet training cannot substitute for work that no longer requires judgment. Judgment isn't missing information. It's missing demand.

In leadership programs, I often hear the same reflection from participants: the training made sense in the classroom, but it didn't prepare them for the reality of the floor. New hires didn't misunderstand expectations; they just hadn't lived them. Long shifts, real-time tradeoffs where multiple priorities collide, and someone had to make a call without perfect information. Information was delivered. Training was completed successfully. But readiness didn't follow because readiness comes from repeated exposure to situations where judgment matters, not from knowing it should matter in theory.

Judgment isn't something people forget how to do. It's not a skill gap that training can close or a motivation issue that better messaging can fix. Judgment is a byproduct of work that requires interpretation. It grows when people are expected to navigate uncertainty, notice subtle changes, and make decisions before there's proof. When the job stops asking for those things, capacity atrophies, even in capable, motivated, well-trained teams. No program overrides that reality.

What the System Is Actually Teaching

An operations leader once said to me, "We keep rolling out programs to fix people, but nobody ever asks what the system is training them to do every day."

That's the issue. The system is always teaching, whether you intend it to or not. Through alerts and workflows. Through escalation paths. Through what gets attention and what gets ignored. Through what creates friction and what lets the day move along. Those signals teach people far more effectively than any workshop or town hall ever will.

If the system teaches compliance, programs that encourage initiative won't change behavior. If the system teaches deference to algorithms, reminders to "think critically" won't stick. The work is the training. Everything else is commentary.

This is why so many improvement efforts stall. Not because leaders don't care, and not because people resist change, but because behavior is addressed while the system that produces it remains untouched. So before launching another empowerment initiative or critical thinking module, pause and ask a harder question: What is the work itself teaching people right now? What behaviors does success require on a normal day, not what your values statement says you want?

If there's a gap between those two answers, training won't close it. Only redesign will.

Why "Speaking Up" Isn't Enough

When judgment starts to fade, leaders often respond with messaging. Speak up. Be proactive. Challenge assumptions. Don't wait for permission. Trust your instincts. The intent is genuine. Leaders want earlier action and better thinking.

But speaking up works if there's somewhere for it to land.

If systems decide before people are expected to think, voice becomes symbolic rather than functional. If uncertainty slows work or attracts scrutiny, people learn to wait. If instinct is treated as noise until data confirms it, instinct stops surfacing. And when early intervention creates friction while late response is procedural, people adjust accordingly.

Judgment rarely fails in dramatic moments. It becomes unavailable at the exact moment it's needed. People still sense when something isn't right, even when the data looks fine. What disappears is the belief that acting on that sense will matter.

And when intuition and the system collide, the system wins. Every time.

> **"We all want to categorize people. We need to listen instead."**
>
> *— Dave MacDonald, Manufacturing Executive, Better Together Group*

The Timing Mismatch No One Sees

Judgment operates upstream of certainty. Before thresholds are crossed. Before deviations are detected. Before escalation can be formally justified. Before anyone can prove that something is wrong. Most systems, however, are designed to act later, once certainty exists. That timing mismatch cannot be solved with training or communication.

By the time escalation is allowed, the system is already in motion. Commitment has been made. Resources allocated. Schedules set. Momentum built. Judgment that arrives then is not prevention. It is damage control. The cheapest options have already passed unnoticed, which is why every response now feels more expensive, more disruptive, and more visible.

Leaders often ask after a failure, "Why didn't anyone act sooner?" But that question assumes action was available. Often it wasn't. There was nowhere for the signal to go. The system did not legitimize intervention on incomplete information. It did not protect someone slowing momentum based on unease. It did not create space for interpretation before confirmation. Leaders ask that question as if time were neutral, when the system has already spent it.

People didn't wait because they were negligent or disengaged. They waited because acting early, questioning the system, or intervening without proof was not expected, not protected, and sometimes not even possible within the structure they were working inside. That is not a training problem. It is a structural one.

Why Motivation Misses the Point

When people appear disengaged, it is tempting to explain the behavior as a motivation issue. They must have stopped caring. They are not engaged enough. They have lost initiative. This is an easy explanation that puts the problem on individuals instead of systems. It is also usually wrong.

People do not stop thinking because they do not care. They stop thinking because it no longer matters. If success is defined by staying inside the lines, people stay there. If deviation is ignored or quietly penalized, people stop offering it. If the system is presumed correct and experience is treated as opinion, authority becomes clear. Not through fear, but through logic. And if caring were enough, your most experienced operators would not be the quietest people in the room.

This is not disengagement. It is alignment with what the work rewards. People are doing exactly what the system has trained them to do: monitor, escalate, and defer. Those are rational responses to work that no longer requires interpretation, questioning, or independent action. Calling that a motivation problem misses the point.

The problem is not people. The problem is that the work stopped asking them to use judgment, and leadership expected it anyway when conditions changed.

The Trap of "Adding It Back"

Eventually, leaders ask the wrong question: "How do we add judgment back?" That question sounds practical. It is actually a signal that judgment was never protected in the first place.

Judgment is not something you reinstall. It is not a feature you toggle on. It does not return because urgency demands it. Judgment grows where the work requires it, repeatedly, under real conditions. If the work no longer requires it, judgment will not reappear because leadership wants it to.

Rebuilding judgment means redesigning work. Redesigning roles so interpretation is expected, not merely permitted. Redesigning authority so early insight can slow momentum without penalty. Redesigning success so noticing matters as much as executing. Redesigning escalation so uncertainty is addressed before it becomes a crisis.

Programs can raise awareness and communicate intent. They cannot override the daily work experience that teaches people whether their judgment counts. Until the work itself changes, judgment will not return, no matter how much leaders want it to.

Why Retrofit Is Exponentially Harder

There's a reason judgment is easier to preserve than restore. Once work has been designed to operate without it, changing that design is expensive and disruptive. Roles are simplified under the assumption that systems will decide. Workflows are optimized around deference. Authority is centralized because algorithms are treated as more reliable. Success is defined by metrics that don't reward early intervention.

Reversing those assumptions requires more than communication. It means renegotiating authority and reintroducing friction. The tradeoff is not fewer false alarms. It is fewer chances to intervene cheaply. That runs directly against years of efficiency gains. Resistance follows, not out of malice, but because both systems and people have been trained to operate differently.

Retrofit isn't impossible. But it's exponentially harder than design, because the longer judgment has been absent, the memory of how to exercise it fades along with the capability itself.

The Line You Can't Cross Twice

I've heard leaders describe this moment in hindsight as "Nothing changed, but everything did." There is a specific point in most automation journeys that leaders miss, and it rarely announces itself. It's the moment when the system no longer needs people to interpret conditions and instead only needs them to confirm what it has already decided.

Most leaders don't notice that moment because nothing breaks. Performance stays strong. Metrics improve. The operation feels calmer, more predictable, more under control. In fact, many leaders feel relief. Fewer questions. Fewer interruptions. Less debate. Everything appears to be working exactly as intended. But beneath that stability, something fundamental has shifted.

Once that line is crossed, the organization enters a different state. Judgment is no longer being built through daily work; it is being slowly depleted. People are still capable. They still show up. They still care. But the work no longer exercises the capability the organization will eventually need. Interpretation gives way to confirmation. Early intervention gives way to escalation. Judgment becomes something people remember having rather than something they practice.

At that point, retrofit becomes necessary. And retrofit is never as effective as design. You are no longer reinforcing judgment through the work itself. You are asking people to unlearn what the system has taught them for years: that the algorithm decides, that acting early carries risk, and that compliance is safer than interpretation. Unlearning those lessons takes time, and during that time the organization remains exposed, caught between habits that no longer serve it and capabilities that have not yet returned.

That's why this line matters so much. Before it's crossed, preservation is possible. Work can be designed to keep judgment active even as automation scales. After it's crossed, the organization is in recovery mode. Recovery is slower, more uncertain, and often incomplete, because you are rebuilding capacity while still operating under pressure.

Most leaders don't realize they've crossed the line until something breaks in a way the system wasn't designed to handle. By then, urgency exists. But the capability to respond does not.

What Actually Has to Change

If programs don't work, what does? Redesign. Not incremental tweaks, better messaging, or stronger encouragement, but fundamental redesign of how work allocates authority, defines success, and creates space for judgment to operate.

That redesign starts upstream. Authority has to move earlier, before commitment, not after. Judgment must be able to slow momentum, question recommendations, and act on incomplete information without requiring escalation. If intervention is legitimate only once proof exists, judgment will always arrive too late to matter.

Success also has to be redefined. If success is measured only by throughput, uptime, and cost, judgment will continue to be optimized away. Early intervention has to count, even when it finds nothing. Questioning the system has to count, even when the system turns out to be right. Slowing down has to count, even when speed was available. If those behaviors aren't visible in how performance is evaluated, they won't survive.

Friction has to be reintroduced deliberately. Not everywhere, and not as chaos, but in specific places where interpretation matters most. Enough resistance to force thought. Enough ambiguity to prevent purely procedural responses. Enough variation to make people decide rather than confirm.

And that protection must be structural. Acting on weak signals cannot be judged solely by outcomes. People must be able to intervene on uncertainty without being penalized for being wrong. That protection cannot depend on a supportive manager or a strong culture. It has to be built into how decisions are reviewed and how success is assessed.

None of this is easy. All of it is possible. But it requires leaders to stop treating judgment as something they can encourage into existence and start treating it as something they must design for and defend.

Why This Feels Backward to Most Leaders

Leaders are trained to optimize. Remove waste. Reduce variation. Increase speed. Standardize responses. Centralize decisions. Automate what's repeatable. Those instincts are valuable. They've made organizations more efficient, more consistent, and more scalable. They've also created a blind spot.

The same instincts that drive optimization also remove the very conditions judgment needs to survive. Redesigning for judgment feels backward because it introduces friction where smoothness existed, distributes authority where centralization worked, and tolerates variance where standardization succeeded.

This discomfort is not a sign that something is wrong. It's a signal that the work is finally addressing the real problem instead of just treating symptoms. Judgment does not develop in frictionless environments. It does not grow where interpretation has been engineered out. And it does not transfer in systems where compliance is safer than questioning.

Here's the reflection point most leaders avoid: if redesign feels uncomfortable, it's probably because it's challenging the same efficiency gains you were rewarded for delivering. That tension isn't a failure of leadership. It's the cost of choosing resilience alongside efficiency.

What Retrofit Actually Looks Like

Before you can retrofit judgment back into an operation, you have to admit something most organizations resist: the system worked exactly as designed. Judgment wasn't lost by accident. It was traded away through decisions that delivered measurable gains.

Understanding retrofit means tracing those decisions. What was optimized? What friction was removed? What authority was centralized? And then confronting the harder question: Who approved those decisions, celebrated the gains, and benefited from the results?

You did. We all did. Because at the time, it looked like progress.

From there, retrofit becomes specific. Judgment doesn't need to exist everywhere. It needs to live where it matters most: at commitment points, at boundaries, where signals conflict, and where consequences compound. The fastest way to find those places isn't a workshop. It's walking the operation and asking people simple questions: Where do you wish you could pause but can't? Where do you see things the system misses? Where would you intervene if you had permission?

Their answers will tell you exactly where to start, if you are willing to listen.

Then the work has to be redesigned in those places. Authority redistributed. Roles reshaped. Success redefined. Protection formalized. And then comes the hardest part: living with the discomfort long enough for new habits to form.

Early interventions that find nothing wrong have to be treated as success, not waste. Questioning the system has to become normal, not exceptional. Slowing momentum has to be legitimate, not career-limiting. That takes time. Months, sometimes years. During that time, efficiency may dip. Variance may increase. The operation may feel less polished.

That isn't failure. That's judgment returning to places where it belongs.

This is where most retrofit efforts die. Not from lack of understanding, but from lack of nerve. Someone will eventually ask why things feel harder, slower, messier. And if leadership can't answer those questions clearly, if they can't defend the friction long enough for capability to rebuild, the organization will optimize it right back out again.

The Decisive Question

Judgment cannot be retrofitted through programs, communication, or encouragement. It returns only through redesign. Which means the decisive question is not whether your people have judgment. They do. The question is whether the work still requires them to use it.

If the work does not demand judgment, people will not develop it, no matter how much training they receive or how sincerely leaders encourage them to speak up. And if the work has stopped demanding judgment, restoring it is not a program, an initiative, or a cultural reset. It is a redesign of how decisions are made, how success is defined, and how the organization values the ability to respond to what it has not seen before.

That work starts with an uncomfortable admission most leaders avoid: the system we built no longer produces the judgment we now expect from it.

So here's your moment of truth: Can you say that sentence out loud in your next leadership meeting? "The system we built no longer produces the judgment we now expect from it."

If that sentence makes you uncomfortable, good. That's the cost of admission to the next section. You can't fix what you won't name.

Once you see that clearly, not simply acknowledge it in a meeting, the conversation changes. The question is no longer "What should we fix?" It becomes "What is our system teaching people to do every day?" And then the harder follow-up: "Are we willing to change that design, even if it makes us slower, messier, and less impressive on a dashboard?"

Here's the reality leaders eventually face. You can run an operation that looks flawless on paper, or you can run one that is ready for reality. You do not get both. Pretending you do is how organizations end up surprised by problems they trained themselves not to see.

The next step is not belief. It is design.

"I have to understand that they are doing what they learned to do, what they thought was right. They weren't coached. Many of the clients I encounter who are frontline leaders were brought up within the company. I'm the first formal leadership coach. So, we have to reprogram that."

— *Marilyn Rosa-Green, LPC, Founder/CEO, Marilyn Rosa Consulting Solutions, LLC*

CHAPTER ELEVEN:

The Authority Problem

> **"Improving safety improved productivity and reduced turnover. People felt they were working in a safe, healthy environment. They stay on the job, and the longer people stay, the more productive they become."**
>
> *— Joe Ricci, President & CEO, TRSA*

Most organizations still follow a familiar sequence. Automate what you can. Standardize what remains. Train people to follow the system. Trust that they will step up when something unusual happens. For a long time, this worked because systems needed people to interpret what they could not sense or calculate. Humans were the adaptability layer, the bridge between imperfect data, delayed signals, shifting demand, and the messiness of real operations.

That sequence breaks once systems can detect, decide, and act faster than humans. Not suddenly and not with alarms, but silently and without drama. Performance improves, reporting looks cleaner, and the underlying logic collapses without drawing attention to itself. What stops working does so beneath the surface, while everything still appears to be under control.

Your maintenance team knows which machines are going to fail before the sensors do. Your quality techs can spot a bad batch before it tests out of spec. Your operators hear when a line is running wrong. But they've learned that "I can tell" isn't a language your system speaks. So they wait for proof. And proof arrives too late to matter.

By then, the organization has already learned how to behave. People wait for the alert, the threshold, the proof, the escalation. And while everyone waits, the system keeps moving.

If no one asked during implementation where judgment would still be required after go-live, the failure mode is already locked in. You simply will not see it until something happens that the system was never designed to notice.

What Leadership Was Never Taught

Here's what business school never taught you: the questions you need to ask before you deploy AI. Where does judgment need to exist before automation scales? Which decisions require interpretation rather than speed? Where should people intervene early, even if it slows the operation? If you deployed automation before you could answer these questions, you built your plant on a foundation you never inspected.

Only after those questions are answered should technology be deployed. Because once automation is in place, habits form, defaults harden, and authority shifts. Normal becomes whatever the system rewards, and anything not protected by design begins to disappear from the work. Not because people stop caring, but because the work stops requiring it.

Leaders misread this because early automation almost always feels like a clean win. Lines run smoother. Variance tightens. Reporting sharpens. Exceptions decline. Sometimes fewer exceptions mean the process is healthier. Other times it means the system has become better at hiding what it cannot count, and the dashboard will not tell you which is true.

To find out, leaders have to go to the floor and ask questions they rarely ask early enough. What are you noticing that the system no longer tracks? What used to get flagged that now moves through without comment? What concerns did you once raise that now feel pointless because the system decides anyway? The answers reveal whether you strengthened the operation or quietly removed its early warning system.

Think about the last three automation projects you approved. At what point did someone ask where human judgment would live after deployment? If that conversation happened after implementation, or did not happen at all, you did not just deploy a tool. You made a decision about what kind of thinking your organization no longer values.

From Managing People to Designing Thinking

In the Smart Plant™, leadership is no longer about getting the best out of people. It is about shaping the thinking the system allows, and just as importantly, the thinking it erases. That requires questions that sound less like HR and more like engineering. Which decisions must always include a human? Where does slowing down add value rather than waste? What signals should interrupt smooth progress even when the metrics look fine? Which roles are expected to interpret uncertainty instead of simply responding to prompts?

> **"Don't ignore the signs your company is getting out of balance. You can't compensate for a lack of culture with an excellent product."**
>
> *— Tom Hatton, Founder/President, Clean Vapor*

Design becomes a core leadership responsibility because the most consequential failures in Smart Plants™ rarely begin as dramatic events. They begin as a gradual loss of interpretive capacity, hidden inside improvements that make everything look tighter and more controlled.

A useful diagnostic is this: Can you name three decisions in your operation that must involve a human, even if the system could handle them faster? If you cannot answer that immediately, judgment is not being designed. You are assuming it will appear when needed, which is risk disguised as confidence.

Most leaders struggle with this not because they dismiss judgment, but because they have never been required to define where judgment lives as a design requirement rather than a cultural aspiration.

How Systems Teach People What Matters

Judgment does not disappear when something breaks. It gets shaped much earlier, through a series of operational choices that feel practical and reasonable at the time. Leaders decide which decisions are automated by default, which roles can interrupt flow, which signals count as meaningful, and which concerns require proof before being taken seriously.

Once a decision is made without judgment present, judgment does not get invited back later. Scope is set. Budgets are allocated. Authority has already moved. Late judgment is not weaker because people lack capability. It is weaker because the decision space has already narrowed.

Systems are not neutral. They encode what matters, when action is allowed, and which signals are legitimate. A system that always decides first teaches people to wait. A workflow that demands proof teaches people to discount unease. A dashboard that rewards smoothness teaches people to avoid disruption. Over time, these lessons harden into how work gets done, and judgment becomes unnecessary rather than discouraged.

What is not required will not develop, will not transfer, and will not be available when conditions change. This is not a culture problem. It is a governance outcome, whether it was intended or not.

How Judgment Gets Designed Out

Judgment is rarely removed by a single decision. It disappears through accumulation. A manual check disappears because it rarely catches anything. A role is simplified to ease staffing. A conversation is replaced with an alert. A threshold is raised to reduce noise. A workflow is hardened to limit overrides.

Each change makes sense in isolation. Together, they redraw the boundary of thinking. What once required interpretation now requires acknowledgment. What once demanded attention now waits for confirmation. What once relied on pattern recognition becomes "follow the prompt." No one announces that judgment is no longer needed. The work simply stops asking for it.

That is why organizations are surprised by outcomes they technically saw coming. Not because people were careless, but because the system trained them to wait. Waiting for confirmation before acting becomes the safest behavior available, and when following process becomes indistinguishable from good performance, the operation runs smoothly right up until it doesn't.

Where Authority Actually Lives

One question exposes the design truth. Where is someone expected to act before the system tells them to? If the answer is nowhere, judgment is not missing. It has been designed out, and predictable consequences follow. Decision-making compresses upward. Escalation becomes the safest move. Weak signals stop surfacing. People become excellent at compliance and less practiced at interpretation.

Leaders often mislabel this as disengagement or lack of initiative. It is neither. It is alignment. People are responding exactly to what the system rewards. When leaders promote flawless execution inside defined parameters and overlook those who interrupt momentum to raise concerns that do not fit neatly into the dashboard, a lesson is taught. When speed is celebrated and caution is treated as drag, another lesson is taught.

The behavior leaders complain about is often the behavior the system has trained.

The Hard Part Leaders Avoid

The hardest part is not seeing where judgment is missing. The hardest part is admitting *why* it is missing. Many of the decisions that displaced judgment were made for good reasons. Safety. Consistency. Throughput. Standardization. Risk reduction. Those reasons still matter.

The problem is not that leaders valued efficiency. The problem is that efficiency became the only design objective. When efficiency stands alone, judgment becomes optional, and optional disappears the moment someone needs to hit a target or justify a cut.

Ask yourself when you last approved an improvement that explicitly preserved space for human interpretation, even though it made the process less efficient. If you cannot recall one, you have been making the same tradeoff repeatedly without naming it. Your operation is now living with the accumulated consequence of those choices.

When Design Becomes Leadership

There is a line many leadership teams avoid crossing, the admission that judgment loss is not a workforce problem. It is a leadership design decision. Not malicious. Not careless. Cumulative.

Each time efficiency was prioritized without naming its cost. Each time speed was rewarded without protecting interruption. Each time clarity replaced conversation. Each time escalation replaced authority. None of these choices is wrong on its own. Together, they create systems that perform well until they encounter something they were never designed for.

Every judgment gap can be traced back to a decision that was applauded at the time. Someone streamlined a process and removed waste. Someone automated a handoff, reducing errors. Someone tightened thresholds and improved metrics. Each win quietly removed a moment where judgment once mattered, and now the organization is living with the compounded effect of victories that felt unambiguous at the time.

The Design Obligation

At this point, the work changes. Up to now, the task has been to see clearly how Smart Plants™ trade judgment for speed, coherence, and control, and why that trade eventually fails. What comes next is design, not aspirational design, but explicit design.

Design decisions about where authority lives. Decisions about what the system must never erase. Decisions about where humans are required to matter. Decisions that defend judgment when pressure builds to remove it, even when the business case looks compelling.

Once judgment leaves through design, it does not return through effort or intention. It returns only when leadership redesigns the work to require it again. That authority lives nowhere else. Not with HR. Not with operations. Not with the people on the floor who have watched their judgment become irrelevant.

The question is no longer whether judgment matters. Everyone agrees that it does. The real question is whether leadership is willing to govern judgment with the same rigor applied to budgets, schedules, and compliance.

So here's your test: In your next leadership meeting, bring up one decision that's currently advancing on autopilot. Production scheduling. Maintenance deferrals. Quality releases. Doesn't matter which one. Then ask this question out loud: "Where in this process is a human required to interpret before the system commits?"

If the room goes quiet, you've found your answer. Every system already reflects a choice. Every workflow already encodes a priority. And every organization is already being shaped by the decisions leadership made easier than the ones they protected.

The only thing left to decide is whether those choices were intentional. Or whether they are the ones you'll spend the next few years explaining, defending, and wishing you had named earlier.

"That key to everything is really around people training.
That's the weak link."

— *Reeves Smith, Director of Sales, Leapfrog Services*

CHAPTER TWELVE:
Reading Your System

You can map every workflow in your operation and still miss where judgment dies. Because judgment doesn't show up in process maps. It shows up in the two seconds between when someone notices something off and when they decide whether it's worth mentioning. Everyone stays professional. Everyone follows the process. No one moves early.

That delay is not a failure of courage or engagement. It is the visible result of an invisible system that has taught waiting to be the safest move available. This chapter shows you how to make it visible and map exactly where judgment gets blocked before you ever hear about it

Two Systems Are Always Running

Every Smart Plant™ runs two systems at once. There is the visible system: equipment, software, workflows, dashboards, and org charts. Then there is the invisible system: who can interrupt momentum, what counts as legitimate evidence, and when uncertainty becomes "real enough" to justify action.

That second system governs judgment. It determines who is allowed to act early, who must wait, and which concerns are taken seriously. Long before an issue reaches leadership, this invisible system has already shaped what happens next.

Stop Diagnosing Behavior

When judgment erodes, leaders often misdiagnose it as behavior. They describe what they see as lack of initiative, excessive escalation, or weak critical thinking. These are not behavior failures. They are system outcomes.

If the safest professional move is to wait, people will wait. If acting early requires proof that exists only after damage occurs, people will delay. The first diagnostic shift is simple but uncomfortable: Stop asking why people didn't act and start asking where early action was ever supposed to be possible.

Think about the last time you were frustrated that "no one raised this sooner." Where, exactly, would that early action have been legitimate?

Follow Authority, Not Intent

To see how judgment really works, follow functional authority, not stated intent. Ignore posters, values statements, and job descriptions. Watch what actually happens when experience conflicts with data. Notice who can challenge a recommendation, who has standing to name unease, and who is allowed to interrupt momentum without consequences.

Then watch what gets reinforced. If escalation is the only safe move, judgment compresses upward. If proof is required before concern is legitimate, early signals disappear. You don't need surveys to diagnose this. You can see it by observing where permission exists and where it does not.

> **"85% of digital transformations fail because companies don't design them to win."**
>
> *—Brent Kedzierski, Manufacturing Technology Leader, kNot Today*

When judgment has been displaced, the pattern shows up as chokepoints: moments where concern must be translated into proof before it can move, where interruption has no owner, and where escalation replaces interpretation. The central diagnostic question becomes unavoidable: Where in this operation is someone expected to act before the system tells them to?

If the answer is nowhere, judgment hasn't faded. It has been engineered out. The next step is locating exactly where that happened.

The Judgment Trace

You can't fix what you can't see. And right now, you can't see where judgment is dying in your operation – not because you're not looking, but because you're looking at the wrong things. The Judgment Trace is how you make the invisible

visible. This isn't an optional reflection. This is operational forensics. Pick one process that runs smoothly. One with green dashboards and a few exceptions. One you're proud of. Now trace what happens when someone notices something the system doesn't confirm. Follow that signal from first observation to final action—or more likely, to where it dies. Map every chokepoint. Every hesitation. Every place where concern needs to be translated into proof before anyone will listen. That map is what you're going to redesign. Not the dashboards. Not the people. The map.

Start with a recent moment of ambiguity. A small quality drift. A maintenance concern. A supplier variation. A near-miss. One of those "weird days" that didn't become an incident. Reconstruct what happened next, step by step. Ignore titles and org charts. Follow functional authority instead.

Ask specific questions. Where did the first concern appear? Who noticed it first? What language did they use to describe it, and did that language have standing in the system or was it dismissed as subjective? What would they have needed to be taken seriously? Where did the signal stop moving? At what point did it become legitimate enough to act on? Who had the right to slow momentum? Who could override the system without multiple layers of approval?

Do this across multiple shifts and supervisors. Track the same type of concern through different hands. At that point, debates about culture usually stop. Structure becomes visible.

The output is a map of judgment chokepoints: places where early signals cannot be acted on without proof, where escalation replaces interpretation, where interruption has no protected owner, and where people have learned that raising a concern without data is career-limiting.

That map is what you redesign. Not people's attitudes. Not engagement scores. Not willingness to speak up. You redesign permission. Until people have explicit permission to act on what they notice before the system confirms it, no amount of empowerment messaging will change behavior.

Run this trace yourself. Don't delegate it. Walk the floor. Ask the questions. Map the chokepoints. If you cannot name where judgment is blocked, you cannot redesign what is blocking it.

A Judgment Trace in Practice

Here's what a trace reveals when done rigorously. Consider a quality drift that resolves itself before triggering any alarms, the kind operators see often but rarely mention because "it worked itself out."

At 7:15 a.m., an operator on Line 3 notices the extruder temperature creeping toward the upper spec limit. Still in range. The dashboard is green. She adjusts manually and logs nothing.

At 7:45, the next operator rotates in. The temperature pattern isn't mentioned during handoff. The dashboard is green, so stability is assumed.

At 9:20, a quality tech samples the product. Viscosity is at the high end of acceptable. It's noted but not flagged. Spec is spec.

At 11:30, maintenance walks the line. A slight change in motor tone is noticed. A mental note is made. No work order threshold has been crossed.

At 1:15 p.m., the shift supervisor reviews dashboards during lunch. Everything is green. There's a vague sense of unease about Line 3, but nothing concrete enough to raise.

At 3:00, the production scheduler sees output trending slightly low and adjusts tomorrow's plan to compensate, without investigating the cause.

The next morning, a different crew arrives. Temperature is mid-range. The drift is gone. No one knows why it appeared or why it disappeared.

What failed here wasn't attention. It was legitimacy. Plenty of people noticed something. No one had a legitimate path to act on "drift" before it became a threshold. What's missing is a role or rule that requires someone to say, "This is changing. We need to understand it before it decides to matter."

Six Diagnostic Questions for Reading Judgment Erosion

Run several judgment traces and the same chokepoints appear. Six questions surface them quickly.

- Where do signals that should connect remain isolated, leaving patterns noticed locally but never assembled into meaning?
- What kinds of concern have no standing in the system, filtering out qualitative judgment that cannot be formalized?
- How long is the gap between first notice and legitimate action, and where does prevention disappear in that delay?
- What is the real cost of raising a concern that may turn out to be nothing, and who absorbs that risk?
- What happens when someone intervenes and nothing bad occurs, and how is that action interpreted after the fact?
- Which roles require less interpretation than they used to, leaving capability present but unowned?

These questions expose how authority actually flows.

What the Patterns Reveal About Authority

Across operations, the same patterns emerge. Legitimacy comes from thresholds, not concern. Timing is reactive, not preventive. Interruption has no protected owner. That's why culture fixes don't work here. Only redesign changes the operating logic.

Once you see these patterns in one process, you start seeing them everywhere. Maintenance. Quality. Engineering. Scheduling. They are not isolated problems. They are symptoms of a system built to execute efficiently and never redesigned to preserve judgment.

Start small. Ask which single chokepoint is costing you the most interpretive capacity. Redesign permission there first. Trying to fix everything at once is how redesign efforts stall.

The Red Flags Leaders Miss

Judgment rarely disappears in dramatic ways. It retreats behind professionalism. Operations still look calm. People are still doing their jobs. Performance still appears strong. But beneath that surface, certain patterns begin to repeat, and they are easy to miss if you are looking only for failure.

Escalation becomes a story rather than a signal. Concerns surface only after they are wrapped in defensible narratives that make action safe, not early enough to make it useful. Questions that should open inquiry are answered with procedure instead, using the rulebook to close conversation rather than guide thinking. Near-misses are explained away instead of examined, with relief replacing curiosity. Seasoned operators say, "I've seen this before," and are acknowledged politely, but nothing moves unless their experience can be translated into metrics the system already recognizes. Over time, silence increases as dashboards stay green, and leadership reads that quiet as maturity rather than a sign that interruption has become illegitimate.

These signals don't point to disengagement or fear. They point to a system that has trained waiting to look professional. If you find yourself proud that things feel calm, uninterrupted, and under control, pause and ask a harder question: What kind of interruption has your system made unacceptable?

The Map You're Building

By the end of a judgment trace, you should be able to draw a map showing where early signals appear and die, where interpretation is required but unsupported, where authority exists on paper but not in practice, where timing prevents intervention, and where learning stops before it becomes structural.

That map is an infrastructure blueprint, not a culture audit. It shows exactly what to redesign and where. Once you can see the invisible system, you stop blaming people for doing exactly what the operation taught them to do. You stop asking for initiative inside a structure that punishes it.

Most organizations never build this map. They see symptoms, launch programs, and blame people when nothing changes. The map gives you something rare: clarity about where judgment is blocked and what is blocking it. That clarity is your starting point for redesign.

Judgment infrastructure does not maintain itself. It has to be defended. And that responsibility now sits exactly where it belongs.

"I give my teams autonomy, but with that autonomy comes ownership of the outcome and the inputs you're going to execute... and ultimately accountability."

— *Brent Hagan, Chief Supply Chain Officer, Lob*

CHAPTER THIRTEEN:
The Five Requirements

By now, you can see the problem. You know where judgment dies in your operation. You can trace it on paper. You could probably walk me through exactly how a weak signal gets ignored while the dashboard stays green. Congratulations, you have a diagnosis.

Now here's the question that matters: What are you going to do about it? Because seeing clearly doesn't fix anything. It just makes you complicit if you don't act. And once you can see the invisible system, you can't blame your people for waiting. Waiting is what the system trained them to do. The next question is whether leadership is willing to redesign the conditions that made waiting the only available option.

The five requirements that follow aren't suggestions. They're the minimum conditions under which judgment can survive automation. If even one requirement is missing, the others collapse. This is the infrastructure. Everything before this was reconnaissance. Everything after this is endurance. This is where the work begins.

Judgment works only when it has a designated place to operate. When leadership says, "Everyone should use judgment," what often happens in fast, automated environments is that no one owns the responsibility. Speed fills the gap. Defaults decide. Empowerment without structure becomes diffusion. Judgment becomes optional, situational, and easy to bypass the moment it slows momentum.

This is the part most leaders miss. Smart Plants™ don't fail because judgment disappears. They fail because judgment disappears in moments when timing still matters. Leadership's task is not to encourage judgment broadly. It is to anchor judgment precisely, at the decision points where early action preserves options.

> **"50% of people in companies are in positions they're not wired for. They migrate there or companies just hire and hope they swim."**
>
> *— John Ballinger, COO, Clean Vapor*

Most leaders discover this too late, after realizing that by the time concern became legitimate, the only options left were explanation and damage control. The signal was there. Someone noticed. What didn't exist was a structure that allowed action before certainty.

What follows are the five requirements that make judgment operational instead of aspirational. As you read them, keep your map from Chapter Twelve in mind. Each requirement should land on a specific chokepoint you already identified. If it doesn't, that's not a theory problem. It's a design gap.

Requirement One: Judgment Must Arrive Before Certainty

Judgment matters most before certainty exists, before thresholds are crossed, alerts fire, or data accumulates enough proof to justify action. It often shows up as discomfort, misalignment, or a sense that something is drifting even while indicators are green. That is not emotion. That is an experienced pattern recognition surfacing early.

Automated systems are designed to act later, after confirmation, repetition, and statistical confidence. That creates a structural mismatch. When judgment is placed downstream of commitment, it no longer changes outcomes. It explains them.

Across manufacturing operations, judgment consistently changes outcomes in three zones:

At commitment points. Once production is released, schedules fixed, or maintenance deferred, flexibility collapses. Early judgment preserves options. Late judgment narrates why options are gone.

Where signals conflict. Systems resolve contradictions mathematically. Judgment resolves them contextually. When no human is explicitly expected to interpret conflict, the system proceeds with the most convenient answer.

Where consequences compound beyond the local decision. Some choices ripple across safety, capability, suppliers, and resilience, and those ripples rarely appear immediately.

If judgment does not live in these places, it will live nowhere that matters. Think about the last issue that "came out of nowhere." Was it really invisible or did it surface in one of these zones where no one was required to act early?

Requirement Two: Authority Boundaries, Not Discretion

Discretion without boundaries does not produce ownership. It produces guessing. In Smart Plants™, guessing is a personal risk, and people respond to it predictably. They choose the safest professional move: usually, to wait.

Smart Plants™ do not run on discretion. They run on decision boundaries, clear definitions of who can act, when, and based on what kind of signal. Automation removes two conditions discretion once depended on: time and containment. Speed compresses timelines. Scale amplifies consequences. Decisions that once affected a single shift can now cascade across the system.

Without boundaries, judgment becomes a career bet. People are left deciding whether intervention is expected or punished. Waiting begins to look like professionalism. If your supervisors cannot name, immediately and concretely, when they are authorized to override a system recommendation without asking permission, you do not have authority. You have polite paralysis.

Decision boundaries assign authority before outcomes are known, not after failure occurs. They answer four operational questions at each critical point: what the system decides on its own, what requires human interpretation, who owns the judgment at that point, and what protection exists for acting early. When any of those is vague, delay replaces judgment.

If you are relying on "good people" to make the right call without telling them when that call is expected, you are not empowering them. You are leaving them exposed.

> **"We do so many things in our daily lives that become second nature, but we don't always know the whys behind things. Why are we approaching this part at this entry angle? That can get lost in a black and white world."**
>
> — *Jim Ver Woert, Senior Client Executive, SME*

Requirement Three: The Right to Interrupt

Speed is not the enemy. Uninterruptible speed is.

One of the most damaging design flaws in advanced operations is the absence of legitimate interruption. Systems run clean. Variance drops. Exceptions disappear. And no one feels authorized to slow momentum.

In many plants, interruption signals breakdown. Stopping the line feels like admitting incompetence. Pausing momentum looks inefficient. Explaining an interruption that "turned out to be nothing" becomes a cautionary tale. That explanation process is not accountability. It is training. It teaches people exactly when not to interrupt.

Judgment operates in the gap between what systems can confirm and what experienced people can already sense. If interruption carries a cost, judgment becomes dangerous. People wait. The organization gets later.

There is also a discreet consequence. When interruption disappears, development collapses. High-potential employees never practice stopping momentum. They never act without proof. They never build the pattern recognition that experience depends on. Veterans intervene and eventually leave, and their judgment leaves with them.

Interruption is where tacit knowledge transfers. Remove interruption, and experience becomes irrelevant. If you cannot point to a moment in your operation where someone has the explicit right to pause momentum based on drift rather than failure, then interruption is not a right. It is a gamble.

Requirement Four: Protection as Structure

Protecting judgment is not about giving people free rein. It is about preserving the structural conditions that make interpretation possible and defending them against constant pressure to improve efficiency.

Protection requires three commitments:

- Authority to interrupt without outcome-based penalty.
- Roles explicitly designed to interpret ambiguity and conflict, not just respond to alerts.
- Enough friction in the system to prevent purely procedural motion when the situation demands thought.

These are design decisions, not cultural values. If they are not deliberately defended, automation will erase them. Every system trends toward speed and uniformity. Judgment survives only where leadership insists that interpretation has a protected place.

Where in your operation are you relying on courage to do the job that the job structure should be doing?

Requirement Five: Learning Before Outcomes

Most organizations learn after failure. Post-mortems. Root cause analysis. Corrective action. That model assumes the organization will get another chance.

In Smart Plants™, that assumption is risky. Some failures compound faster than learning can catch up. Some outcomes are irreversible. Judgment has to be strengthened before outcomes prove it right or wrong.

That means evaluating early intervention on the quality of reasoning, not just on results. It means treating "false alarms" as practice rather than waste. It means building pattern recognition through exposure, not hindsight. If being wrong is punished more than being late, people will wait for certainty. And certainty arrives too late to matter.

Organizations that learn only after failure depend on failure to teach. Organizations that learn before outcomes treat judgment itself as infrastructure. The difference is not philosophical. It is structural, and it shows up long before the next surprise.

What These Requirements Create

When judgment operates before certainty, with clear authority boundaries, legitimate interruption, structural protection, and learning that doesn't require failure, the organization develops a different kind of intelligence. Not one that simply masters the known, but one that adapts when reality refuses to cooperate with the plan.

What makes this difficult is that erosion happens while everything still looks fine. Metrics stay green. Systems run smoothly. Productivity improves. Leaders often sense something is missing in the room, but they can't point to a number that proves it. By the time the absence becomes visible, the organization has already learned to suppress the signals that would have revealed it earlier.

These five requirements are not the finish line. They are the structural baseline. Without them, every new tool, dashboard, and process will narrow the space where judgment can operate. With them, judgment has a defined place in the system when conditions deviate from what it was designed to handle.

And here is the part that requires honesty. You already know whether these five conditions exist in your operation. You know whether people can act before proof arrives. You know whether interruption is legitimate or subtly punished. You know whether judgment is protected by design or left to personality and courage.

The question is not whether you understand the requirements. The question is whether you are willing to redesign around the one that is most broken.

Do not try to fix all five at once. That is how redesign becomes initiative fatigue. Start with the one requirement that, if strengthened, would return the most judgment capacity to the system. Make it explicit. Protect it. Govern it. Then move to the next.

Here's how you choose which one: Look at your map from Chapter Twelve. Look at the chokepoints you identified. Now ask yourself: Which single requirement, if you fixed it at one of those points, would change behavior in the next thirty days?

Not which would be easiest. Not which would look best on a project plan. Which would change how decisions actually get made when something doesn't fit the model.

That's your starting point. Not because it's strategic. Because it's urgent.

Because once these requirements are in place, something shifts. Not just in performance, but in how the organization thinks. Authority becomes clearer. Signals move earlier. Decisions preserve options rather than defend them.

And that shift is what separates operations that manage stability from operations that can handle surprise.

"You're talking about courage to put that information together and to look at it honestly. And then you're also talking about trusted data. There's data, and then there's trusted data. People were always questioning the data: Whose data is that? Where did you get this information? How do we know it's correct?"

— *Vince Sassano, President, Strategic Performance Company*

CHAPTER FOURTEEN:
What Dashboards Hide

Here's what your board doesn't want to hear: your plant can be hitting every target and dying at the same time. The dashboards? Liars dressed in green. They're showing you what happened yesterday, while tomorrow is silently breaking down.

Smart Plants™ do not usually collapse in a blaze of alarms or cascading system failures. They fail while appearing competent. They fail while metrics improve. They fail while leaders are congratulating teams for fewer interruptions, fewer escalations, and fewer deviations from the plan.

That is why this problem survives. It hides inside progress.

When interventions decline, leaders assume the system is healthier. When escalations drop, they assume the organization is maturing. When the plant runs smoothly, they assume it is safer and that judgment has finally done its job.

Most leaders can point to a meeting like this: A dashboard review where everything looked better than last quarter. Fewer alerts. Fewer interruptions. Fewer "unnecessary" escalations. The progress felt earned. No one in the room asked the harder question: Are we seeing less risk, or are we seeing less of what used to make risk visible?

Sometimes the system is healthier. And sometimes the organization is learning a different lesson entirely. It is learning what is safe to say, what is expensive to mention, and which kinds of thinking no longer have a place to land.

Metrics Don't Reflect Reality. They Define It.

Leaders often speak about metrics as if they are neutral mirrors. Objective. Factual. Detached from interpretation. But the moment a metric exists, it legitimizes attention. It authorizes concern. It gives someone permission to act.

What does not exist as a metric becomes harder to name, harder to defend, and easier to dismiss as opinion or noise. Authority follows what is counted.

In a Smart Plant™, metrics are not just a scoreboard. They are a language of legitimacy. They determine which realities are allowed into the room and which must fight to be heard. If a condition appears on a dashboard, it becomes real. If it does not, it becomes anecdotal. If it cannot be quantified, it becomes subjective. If it does not cross a threshold, it becomes "not actionable."

The signals that save operations rarely arrive in the tidy form leaders prefer. They arrive early, carrying ambiguity. They show up as discomfort that has not yet been formalized. They appear as patterns that someone can feel before they can prove.

Metrics resist ambiguity, and so do most executive teams. Organizations build systems that reward certainty and call it discipline, then act surprised when the people closest to the work stop speaking until certainty arrives.

If your dashboards are the only language of legitimacy in your operation, ask yourself what kinds of risk no longer have a microphone.

Why Dashboards Feel So Reassuring

Dashboards are seductive for reasons most leaders rarely admit. They simplify complexity. They compress uncertainty into color. Green signals acceptable. Yellow signals caution. Red signals action. That structure reduces debate and limits interpretation. It offers something leaders crave in complex environments: relief, the sense that ambiguity has been managed and control restored.

The danger is not inaccuracy. The danger is direction. Dashboards are built to track what organizations already understand: throughput, efficiency, variance reduction, known failure modes, and historical comparison. They are precise within those boundaries.

What they are not built to capture is early drift, slow degradation, contextual unease, cross-system interaction, or emerging risk. What does not appear on the screen begins to feel less legitimate. And when something feels illegitimate, it gets filtered out – first socially, then structurally, and eventually automatically.

Once smoothness becomes synonymous with safety, interruption begins to look like interference.

The Signal-to-Noise Myth

Few phrases do more quiet damage than this one: "We raised the threshold to reduce noise." On the surface, it sounds disciplined and efficient. It suggests focus and maturity. In practice, it often means early signals were removed because they were inconvenient, ambiguous, or difficult to interpret.

Noise is rarely noise to the people closest to the work. More often, it is uncertainty that has not yet been formalized. It is pattern recognition before validation, a weak signal that does not fit neatly into the categories the system was designed to recognize. When those signals are filtered out, nothing looks broken. The dashboards calm down. The alerts decrease. Conversations get shorter.

But when thresholds rise, risk does not disappear. It shifts downstream to a point where options are narrower and response costs are higher. The organization grows calmer, and leaders mistake that calm for stability. In reality, the system has simply postponed its discomfort.

This is how judgment gets engineered out while leadership believes it is engineering excellence in. The numbers improve. Friction fades. Variance declines. And the organization loses time without realizing it.

If your system requires stronger proof each year to justify early action, pause and ask yourself a harder question: What kind of signal no longer qualifies as worthy of attention, and who decided that standard was necessary?

How Metrics Train Compliance

No leader sets out to suppress judgment. But metrics reward very specific behaviors, whether intended or not. Over time, people learn what earns recognition and what creates friction. They learn to wait for confirmation. They learn to translate concerns into approved language. They learn to escalate rather than interpret. They learn to align with the dashboard rather than with what they are sensing on the floor.

People become excellent at monitoring, acknowledging, and reporting. They become less practiced at noticing, interpreting, and intervening. Not because they stopped caring, but because the system no longer requires that capability to function until it suddenly does.

This is why the shift is so hard to see. The organization still looks disciplined. The plant still hits its numbers. The system is delivering exactly what it promised.

What changes is the ratio between two kinds of work: work that executes and work that makes sense of what execution cannot explain. Metrics are excellent at rewarding execution. They are indifferent to sensemaking. So sensemaking becomes unpaid labor – the thinking people do quietly if they have time, feel safe, and believe anyone wants to hear it. If not, it disappears.

What Metrics Silence First

Metrics rarely silence everything. What they silence are specific categories of information, the kinds of signals that do not fit neatly into structured reporting. Weak signals that cross functional boundaries. Early pattern recognition that lacks clean attribution. Experiential knowledge that resists quantification. Concerns that feel real but arrive before proof.

Those signals do not disappear. People still notice them. They simply learn that there is nowhere legitimate to put them. When a signal has no sanctioned destination, it does not vanish. It lingers in side conversations, in gut feelings, in passing remarks that never make it into a meeting summary.

Eventually, a signal resurfaces as an incident, a surprise, an "unexpected" failure that many people quietly anticipated. That is how organizations become efficient and blindsided at the same time.

When Metrics Become Authority

Here's the shift you need to understand: the moment metrics stopped informing your decisions and started making them. When was the last time someone challenged the dashboard in your leadership meeting and won? When was the last time experience trumped data? If you can't remember, your metrics aren't tools anymore. They're your boss. And they're a terrible one.

When the dashboard becomes the final word, judgment is forced downstream. Intervention becomes escalation. Thinking becomes justification.

Leadership teams then fall into a predictable trap. They say they want early judgment but reward late certainty. They call for initiative but compensate compliance. They ask why no one acted sooner while building systems that penalize acting without proof.

That tension is not cultural. It is structural. Authority has migrated into measurement, and once authority is embedded in tooling, it becomes difficult to challenge. Questioning the metric begins to feel like questioning reality itself, even when everyone on the floor knows something important is missing.

In your last leadership meeting, how many decisions were shaped by interpretation, and how many were closed by the numbers alone? When someone says, "The data doesn't support that," does the conversation deepen or end?

When numbers end the discussion rather than inform it, authority has shifted. And measurement systems optimize for what they can see, not for what the operation will need next.

The Plant That Stops Learning While It Improves

Smart Plants™ can improve while learning slows down. That sentence makes most executives uncomfortable because it challenges a deeply held belief that better systems automatically produce better understanding. In reality, improvement and learning are not the same thing, and in highly optimized environments, they often move in opposite directions.

Measurement answers historical questions. It tells you what happened, how often, and how much. Sensemaking answers directional questions. It asks what this means, why it is happening now, and what might be changing beneath the surface. Metrics are excellent at tracking history. Judgment enables an organization to anticipate direction. Dashboards can show you where you have been. They cannot tell you what you are drifting toward.

Drift is the real story in most operational failures. Not a dramatic breakdown, but a gradual shift. A process that slowly changes shape. A supplier that quietly degrades. A maintenance plan that begins deferring what it used to catch. A training pipeline that narrows year after year while productivity holds steady. The performance story stays polished. The underlying reality shifts.

When measurement improves without protecting sensemaking, learning becomes reactive. Insight arrives only after consequence. The typical response is to tighten the system further: more KPIs, narrower thresholds, additional alerts, and sharper reporting. The operation becomes more analytically sophisticated. It also becomes more dependent on proof.

I have watched this pattern repeat across industries. Leaders point to their dashboards and say, "We are more data-driven than ever." And they are correct. What they often miss is that relying heavily on data and being judgment-capable are not the same thing. Data tells you what already occurred. Judgment determines what to do before the next occurrence hardens into cost. If an operation strengthens the first while weakening the second, it does not become smarter. It becomes better at measuring its own decline.

Metric Governance as Infrastructure

If judgment is infrastructure, then metrics are not passive instruments. They are structural components of authority. They allocate timing. They define permission. They determine when action is legitimate and when unease must wait.

That makes metric governance a leadership responsibility, not a reporting exercise.

Every metric shapes thinking. It signals what deserves attention and what can be ignored. It tells people whether interpretation is expected or whether the system has already decided. If you do not consciously govern those effects, the organization will drift toward speed and friction reduction by default.

The first governance question is deceptively simple: What can your system not afford to silence?

Every operation contains signals that are operationally vital and structurally difficult to measure. They are ambiguous. They cross boundaries. They show up early. Leaders often label them as soft. In practice, they are frequently the earliest indicators of real risk.

Consider the kinds of signals that rarely trip a red indicator: a slight but scattered rise in rework, minor safety events that feel related but remain below reporting thresholds, increasing reliance on one experienced technician to stabilize complexity, new hires who can execute tasks but cannot explain cause and effect, and maintenance teams closing tickets faster because deeper diagnosis has been deprioritized. None of these may look urgent. All of them may be telling you something important.

If your measurement system provides no legitimate landing place for these signals, your plant will not see them in time. People will still notice. They will simply stop raising them.

A second governance decision involves distinguishing between metrics that inform judgment and metrics that replace it. Some measures are descriptive and provide context for interpretation. Others function as gates: green authorizes movement and red blocks it. That may be appropriate in highly standardized domains. Gate-based metrics become dangerous when extended quietly into areas that still require interpretation.

If leadership does not explicitly draw that line, the organization will collapse the distinction in favor of reduced friction. Reduced friction feels efficient. It also removes the moments where human judgment once added range.

A third decision requires creating legitimate room for disagreement with the system. High-caliber operations build pathways that allow interpretation to challenge outputs before proof is complete. Most plants do the opposite. They require confirmation before concern is allowed to matter. That turns early judgment into personal risk and gradually trains waiting as professionalism.

When disagreement lacks a structured channel, interruptions decline. The organization appears calmer. It may simply be later.

Finally, every metric system embeds a timing model. It communicates when intervention is allowed. Some encourage acting on weak signals. Most reward waiting for confirmation. If action is legitimate only after a threshold, intervention is late by design. If escalation is required, judgment is centralized. If proof is required, hesitation is trained in.

These are not accidental outcomes. They are architectural choices.

The Decision You're Actually Making

The real decision is not whether to measure. Every serious operation measures. The decision is whether your measurement system preserves the ability to notice what has not yet stabilized or gradually conditions people to move past it in the name of discipline and efficiency.

Here's your diagnostic: Pull up your dashboard right now. Look at the metrics that drive your leadership meetings. For each one, ask yourself: When was the last time someone challenged this number based on what they were sensing on the floor? When was the last time experience won an argument against data?

If you can't remember, your metrics aren't tools anymore. They're the authority. And they're making decisions you don't even realize you've delegated to them.

What erodes first is not performance. It is range. The range to handle what the system was not explicitly designed for. The range to detect drift before it becomes visible in cost or failure. The range to act on informed concern rather than documented proof.

Most leaders do not need an audit to know whether that range still exists. They can feel it in the tone of meetings. They can see it in whether data closes conversations or deepens them.

If your most influential metrics function primarily as gates rather than guides, your system is not simply measuring performance. It is defining when judgment is allowed to operate. And by the time your dashboards clearly display a problem, you are no longer preventing it. You are recording it.

That is the trade most Smart Plants™ never intend to make. They gain control and lose range. They gain precision and lose perspective. They become better at explaining what happened and weaker at shaping what happens next.

The question is: Are you willing to measure differently before you're forced to explain why you didn't?

"I think it's a shame that we want to force people to fit into a box. As a person of color, I get the code-switching. I grew up learning to do it. It's exhausting to pretend that you fit into a place where you really aren't able to be your full self."

— *Karla Trotman, President and CEO, Electro Soft Inc.*

CHAPTER FIFTEEN:

Judgment as Requirement

Chapter Fourteen made one thing clear: Metrics do not simply measure performance; they define legitimacy. Once authority migrates into dashboards and thresholds, judgment becomes discretionary. And discretionary capability does not survive acceleration.

If judgment has been engineered out, leadership must deliberately redesign it. No one else holds that authority. The moment leadership accepts that ownership, the conversation changes. This is no longer about awareness or cultural encouragement. It becomes about specification.

A Smart Plant™, as defined in this book, is not simply a factory with more software layered on top. It is an operation in which sensing, decision-making, and action have been accelerated by automation, analytics, and AI. When decisions move faster than people, the system begins determining what gets noticed, what gets acted on, and what gets ignored. Speed reshapes authority. And unless judgment is explicitly required, it gradually becomes optional.

Every Smart Plant™ already specifies what it depends on. Uptime requirements are defined. Safety tolerances are codified. Quality thresholds are explicit. Capital approval limits, cybersecurity protocols, financial controls, and escalation paths all are treated as structural requirements. The system assumes they exist. Work is designed around them. Contingency planning accounts for them. When they are absent, the operation is considered incomplete or unsafe.

Judgment is almost never treated this way.

When the system does not require judgment, the people who carry that capability compensate instead. They translate concerns into acceptable language. They soften warnings to reduce friction. They absorb personal risk to prevent something worse from happening. That work is invisible, and because it is invisible, it is easy to assume it is unnecessary. Over time, it becomes exhausting. Eventually, it disappears – not because people lose capability, but because the cost of compensating for structural gaps becomes unsustainable.

Most operations are not short on people with good judgment. They lack a system-level dependency on it.

Think about your most experienced operators. They make dozens of small judgment calls every week that prevent issues the system would never detect. They adjust before thresholds trip. They notice patterns that dashboards cannot represent. They intervene without documenting the activity because the correction feels like common sense. When the line runs smoothly, leadership sees proof that the system works. Leadership rarely sees proof that judgment just protected the operation. The next round of optimization removes a little more margin, a little more discretion, a little more space. Eventually, those operators stop compensating. Not because they cannot, but because the system no longer requires their input.

If you pause for a moment, you can probably name where that margin has already narrowed in your operation.

Judgment as a Missing Dependency

In engineering terms, a dependency is something without which a system cannot function safely. When a dependency is missing, the system may still operate, but it borrows stability from elsewhere, usually through informal, fragile compensations.

That is how judgment survives in many advanced operations today. An experienced supervisor pauses a line without explicit authority and explains it afterward. A maintenance lead overrides a recommendation based on instinct and takes on personal exposure. A quality manager pushes back on a release decision even though every indicator is green.

These actions are not evidence of undisciplined behavior. They are signals that the system is incomplete.

When judgment exists only as informal compensation, it becomes dependent on tenure, credibility, courage, and relationships. It does not scale. It does not transfer cleanly. It disappears when the individuals carrying it leave, burn out, or decide that absorbing personal risk is no longer part of their job description.

Leadership often misreads this pattern as resilience. It is debt. And like all debt, it looks manageable until conditions change.

Why Encouragement Is Not Enough

Most organizations attempt to preserve judgment through encouragement. Leaders urge people to speak up. They promote empowerment. They reward critical thinking in performance reviews. None of this is misguided. It simply collides with how systems behave under speed.

Encouragement relies on discretion. Discretion collapses when timelines compress, ambiguity increases, and consequences scale. Under pressure, people do not act based on encouragement. They act in alignment with what the system requires or punishes.

If workflows advance automatically, people learn to stay out of the way. If metrics define legitimacy, people align with the metric. If proof is required before intervention is recognized, people wait for proof. Encouragement becomes background language while structural incentives shape behavior.

Consider the last time someone in your organization stopped a process based on informed concern and turned out to be wrong. What happened next? Was that action treated as responsible judgment exercised in good faith, or as a disruption that should have waited for more data? The answer to that question tells you more about your real operating logic than any value statement.

People are not confused about what matters. They observe what is reinforced and adjust accordingly.

Specification Changes the Work

Treating judgment as a requirement shifts the leadership conversation from motivation to mechanics. The questions change. Where is human interpretation required for this operation to remain stable? Which processes must pause unless a human has weighed in? Which decisions cannot proceed without a named owner of judgment? Where does acting early matter more than being right?

These are architectural questions. They expose tradeoffs that leadership often prefers to leave implicit. Every requirement introduces friction somewhere. Every protected human intervention slows down something. Every pause challenges the assumption that speed always equals progress.

Yet Smart Plants™ already accept this logic in other domains. No one argues that safety interlocks are optional. No executive treats financial controls as discretionary. No board considers cybersecurity a preference rather than a requirement. Judgment protects against a class of risk that no automated system can fully anticipate: novelty.

And novelty is no longer an exception. Supply chains shift. Regulations evolve. Market conditions change. Customer expectations move faster than capital cycles. The idea that every scenario can be anticipated and encoded is an expensive form of optimism. Judgment bridges the gap between what your system was designed to handle and what reality delivers. Treat it as optional, and you are betting that nothing outside your design assumptions will occur.

That is not a strategy. It is hope disguised as confidence.

What It Means to Specify Judgment

Specifying judgment does not mean inserting humans everywhere. It means deciding where the system must not be allowed to proceed on its own.

Across plants, the same zones consistently emerge. Decisions that create commitments that are hard to reverse. Situations where signals conflict and rule-based logic resolves the conflict even when context suggests otherwise. Local optimizations that quietly generate non-local consequences affecting safety margins, supplier resilience, capability development, or future flexibility.

If judgment is not required in those zones, it will not appear when it matters.

Many leaders believe they have already designed for judgment because intervention is technically possible. A supervisor can override the system. An engineer can escalate concerns. A manager can call a meeting. But access is not the same as obligation. If the system can proceed without human involvement, intervention becomes discretionary. Discretion becomes personal exposure. Exposure becomes hesitation.

Specification reverses the burden. When judgment is required, the system waits. Momentum pauses. The absence of interpretation becomes visible. Action without human review becomes the exception rather than the default.

Empowerment grants permission. Specification creates necessity. Necessity sustains capability.

Why This Belongs to Leadership

No one downstream can do this work for leadership. Engineers optimize within defined constraints. Operators execute what is specified. Managers enforce consistency. Only leadership can decide where speed must yield to sensemaking and where automation must pause for interpretation.

Every system operates on assumptions frozen in time. Judgment is what allows an organization to notice when those assumptions begin to erode. If leadership does not specify where that noticing must occur, the system will optimize past it.

This is not about distrusting your systems or slowing your operation indiscriminately. It is about recognizing that the more capable your systems become at handling what they were designed for, the more critical it becomes to preserve the capability that handles what they were not. That capability does not sustain itself through good intentions or occasional reinforcement. It survives only when the system cannot function without it.

Once judgment is specified as a requirement, leadership must decide who holds it, when it acts, how it is protected, and how it is evaluated. That work is not a policy statement. It is ongoing governance. It requires leaders to ask regularly where judgment operated, where it was needed but absent, and where the system moved forward without it because no one was required to pause. If you cannot answer those questions clearly, judgment is still a preference rather than a dependency.

Let me make this concrete: You require three signatures to approve a $25,000 purchase. You require safety training before anyone touches a machine. You require financial controls on every transaction.

Where do you require judgment? Not encourage it. Not value it. Require it.

If you can't point to a place in your operation where the system cannot proceed without human interpretation, you haven't required it. You've wished for it. There's a difference.

Eventually, experienced operators will stop compensating, not because they do not care, but because no one can carry invisible infrastructure indefinitely. When that compensation disappears, the failure will look sudden. It will be described as a training issue or a communication breakdown. In reality, it will be the delayed cost of a requirement you never specified.

Chapter Sixteen moves from requirement to structure. Because once judgment becomes mandatory, the organization itself must be designed to sustain it.

"What went right? What went wrong? What did we learn? And what are we going to do in the future as a result?"

— *Brent Hagan, Chief Supply Chain Officer, Lob*

CHAPTER SIXTEEN:

The Operating Architecture

Once judgment becomes a requirement, three questions become unavoidable: Who holds the authority to act? When is that action legitimate? How does the organization learn from it?

These are not separate problems. Together, they form an operating architecture. When aligned, they create conditions where judgment survives optimization. When misaligned, they produce fragility that looks like discipline. Most organizations have never consciously designed this architecture. It emerged through accumulated operational decisions that felt tactical at the time but were structural in effect. And what emerges by default in automated environments always privileges momentum over interpretation unless leadership intervenes.

In Chapter Fifteen, you accepted that judgment must become a system requirement. That decision changes the conversation. But declaring that judgment matters and building a structure where it can operate are very different acts of leadership. Many organizations skip from "We need more judgment" to "Why aren't people using judgment?" without ever examining whether the system gives it a legitimate place to live. This chapter closes that gap.

Authority Before the System Decides

Once judgment is required, it still has to belong to someone. And in many plants accelerating toward automation, authority has quietly migrated to the wrong place.

On paper, authority appears orderly. Titles are defined. Approval limits are documented. Escalation paths are mapped. In practice, authority lives where decisions are finalized without friction, where momentum is hard to interrupt, and where commitments are costly to reverse. Increasingly, that point sits downstream of automation. Systems sequence work before supervisors engage. Algorithms prioritize before planners weigh tradeoffs. Dashboards legitimize action before context is discussed. Humans are involved, but later, after confirmation, after escalation, after options narrow.

That shift is often described as efficiency. It is also a relocation of authority. Authority no longer moves simply up or down the org chart; it moves forward into the system itself.

Walk into your operation tomorrow morning and identify three decisions that will be made before 10 a.m. For each, ask at what moment does a human have the authority to say, "Wait. Let's think about this differently." If the answer is after resources are committed, after schedules are locked, or after escalating through multiple layers, authority has already moved past the point where judgment could meaningfully shape the outcome. You are not governing the decision. You are ratifying what the system has already set in motion. And ratifying inevitability is not leadership. It is supervision after the fact.

Late authority feels controlled. The data appears more definitive. Alternatives are fewer. Accountability seems clearer because the system has already framed the decision boundaries. But comfort comes at a cost. When authority arrives after commitment, it no longer shapes direction. It manages consequences. Leaders become reviewers of inevitability rather than designers of trajectory.

Escalation is often mistaken for proof that authority still exists. If something feels wrong, escalate. If uncertainty remains, escalate. If the recommendation seems off, escalate. When escalation becomes the only legitimate response, judgment compresses upward. Frontline teams stop interpreting and start packaging. Middle managers optimize defensibility. Senior leaders arbitrate exceptions far removed from the original signals. Everyone remains busy. Almost no one is early.

If your senior team spends most of its time resolving edge cases, that is not a sign of engagement. It is a signal that authority has been placed downstream of commitment.

Authority in Smart Plants™ is not primarily about rank. It is about timing. Who can act before the system commits? Who can slow momentum without penalty? Who can decide with incomplete information? If those rights activate only after escalation, authority is late by design. And late authority will always favor continuity over correction.

Resilient plants do not distribute authority evenly. They deliberately place it at the moment of commitment, where signals first conflict and where local optimization creates irreversible downstream effects. Authority that exists only on paper does not shape behavior. Authority that exists at the moment of decision does.

When authority is unclear, organizations do not become chaotic. They become careful. People align with the system because it is safer than interpretation. After failure, leadership asks a familiar question: Who approved this? The answer is usually procedural. The rule was followed. The system recommended it. The data supported it. No one owned the judgment.

> **"We don't hire people. We select people for our team, much like a professional team does."**
>
> — *John Ballinger, COO, Clean Vapor*

That is not a people problem. It is a governance failure.

Timing Is the Hidden Architecture

Authority answers who can act. Judgment shapes how decisions are made. Neither matters if timing makes early action illegitimate.

Every system embeds a theory of time. Some encourage intervention on uncertainty. Most reward waiting for confirmation. In highly automated environments, the default timing model is simple: act after thresholds are crossed, after alerts fire, after escalation justifies intervention. It feels responsible. It feels disciplined. It is also how operations get faster and less capable at the same time.

Proof narrows interpretation and reduces debate. But by the time proof exists, flexibility has often disappeared. Materials are committed. Schedules are locked. Resources are consumed. Judgment at that stage manages cost; it rarely prevents it. Timing is not about speed. It is about whether judgment is allowed to operate while options still exist.

Timing is rarely designed directly. It emerges through thresholds, alerts, escalation protocols, and workflow gates. Each embeds an assumption about when action becomes legitimate. Over time, the organization learns the rule: early action is unnecessary, sometimes risky, and often discouraged. So people wait. Not because they are disengaged, but because they understand the timing rules better than leadership does.

Early action is not premature action. Acting while options remain open is fundamentally different from acting recklessly; the distinction is structural, not emotional. When early intervention operates within defined ownership, bounded scope, and protected intent, it produces learning instead of chaos. When those guardrails are absent, intervention feels like personal exposure rather than responsible stewardship. If early action in your plant is routinely mistaken for overreaction, the issue is not the individuals stepping forward. The issue is that the design never established the difference between recklessness and a disciplined pause.

There is a narrow window where judgment creates the greatest leverage: after noticing, before confirmation; after concern, before commitment. Most systems skip that window. They move from "not actionable" directly to "urgent." They leave no sanctioned space for interpretation. That gap is where resilience erodes.

Senior leaders can attend every review and study every dashboard and still feel late. That is not inattentiveness. It is a timing design. If legitimacy activates only after escalation, leadership will always be reacting to consequences rather than shaping conditions. Timing is not an operational detail. It is governance.

Designing Early Legitimacy

Resilient Smart Plants™ do not require people to be right early. They require people to act early in bounded ways. Early action pauses momentum rather than reversing it. It creates space for interpretation rather than forcing resolution. Most importantly, it is legitimate.

When early action is legitimate, weak signals surface sooner, and discomfort becomes usable information. When it is not, judgment remains private until it becomes provable or irrelevant.

Authority and timing, however, are still insufficient if the organization cannot learn from what early judgment produces.

Learning Before Outcomes

Most organizations are built to learn from failure. Incidents trigger reviews. Breakdowns generate analysis. Near-misses produce reports. These mechanisms are familiar and legitimate. Early judgment disrupts that model because its purpose is to prevent the outcome from occurring. When it works, there is no incident to study.

Without deliberate design, early intervention is reclassified as inefficiency. The pause looks unnecessary. The concern appears overcautious. The lesson becomes: next time, wait. That quiet reinterpretation is how judgment retreats.

Outcome-based learning fails early judgment because it evaluates correctness rather than reasoning. "Was this right?" judges the person. "What did this reveal?" strengthens the system. If review meetings revolve around being right, you are training caution rather than capability.

Early action generates information that dashboards cannot: where assumptions are beginning to erode, which signals matter before they are measurable, where authority and timing strain under pressure, and whether protection is real or rhetorical. If that information is not captured deliberately, it evaporates.

Learning from early judgment requires memory.

Capture the context of the decision: What type of commitment was involved? Who exercised authority? How far upstream did they act?

Capture the trigger: What pattern or signal surfaced concern before proof existed?

Capture the reasoning: What interpretation led to the pause? What alternatives were considered? What experience informed the decision?

Capture what was discovered during the pause: Were authority and protection held under scrutiny?

Capture what changed structurally as a result: Did decision boundaries shift? Did monitoring parameters adjust? Did training content evolve?

If you document only failure, you will teach only failure. If you document early judgment, you build transferable intelligence. Without that memory, capability stays personal and leaves when the person leaves.

Over time, patterns accumulate. Certain signal combinations recur as reliable early warnings. Certain authority placements prove too constrained or too loose. Certain timing windows consistently prove late. This accumulated intelligence allows the architecture to be refined based on lived practice rather than assumptions. Authority can be redistributed based on evidence. Timing windows can move earlier because experience supports it. Protection mechanisms can be strengthened to address outcome bias that still creeps in.

Learning itself is a timing decision. It must begin while ambiguity remains. Once certainty arrives, the story simplifies, and competing interpretations disappear. Resilient plants treat early interventions as learning events, not interruptions. They do not ask only whether the intervention was correct. They ask what it revealed about the system's assumptions.

An organization can survive early action without learning from it. The line runs. The metrics remain stable. The crisis never appears. But survival is not adaptation. Adaptation requires memory: remembering not just what happened, but what almost happened and why.

When early judgment feeds learning, the system evolves. When it does not, the system resets. And systems that reset relearn the same lessons later at a higher cost.

Alignment Across Decision Types

Authority placement, timing legitimacy, and learning discipline are not abstract concepts. They show up at concrete commitment points: material release, maintenance scheduling, production rate changes, supplier transitions, safety procedure modifications, and staffing decisions. In each case, the same pattern tends to repeat. Authority migrates to systems or distant roles. Timing defaults to threshold-based action. Early signals have no legitimate pathway to reshape decisions. People closest to the consequence lose influence as speed increases.

Each of these shifts feels reasonable in isolation. Moving authority into systems reduces variability. Requiring proof before action reduces perceived overreaction. Optimizing schedules improves cost metrics. But stack these reasonable adjustments across every decision type and you have systematically relocated judgment away from the moments where it would have preserved optionality. What looks like six small efficiency gains becomes one large capability loss.

If you ask three leaders the same questions – "Who can stop this? How early? Based on what?" – and receive three different answers, the architecture is not flexible. It is undefined. Ambiguity here does not create empowerment. It creates a delay.

For each critical decision type, four questions must have explicit operational answers. Where does commitment truly occur – not administratively, but practically? Who can act before that commitment without escalation? What patterns or signals make early action legitimate before proof exists? How is early action evaluated – by the quality of reasoning and the options preserved, not by whether the outcome justified it in hindsight?

If those answers are vague, the architecture is aspirational rather than structural. And aspiration collapses under pressure.

Designing Is Not Defending

By this point, you have a blueprint: authority placed upstream, timing that legitimizes early intervention, and learning that captures what judgment reveals. Designing that architecture is intellectually satisfying. Defending it is politically difficult.

Because the pressure to optimize never stops. Someone will argue that an "unnecessary" pause slowed production. A high performer will resist bounded authority because it feels like friction. Quarterly targets will make early intervention look inefficient in hindsight.

Here's what you need to be ready for: The first time someone exercises the authority you just designed, and nothing bad happens, someone in your leadership team will say, "See? That pause wasn't necessary." That moment is your test.

If you defend the reasoning anyway, you've built infrastructure. If you allow the question to linger, you've built theater.

The architecture you have just designed will not erode due to external disruptions. It will erode because someone makes a compelling case that it's slowing down the business. And if you're not ready to defend it in that moment, it was never infrastructure. It was an experiment that just failed.

Designing the operating architecture is the first act of leadership. Defending it when it's tested is the second.

"People will do what they committed to if you don't sabotage or betray them, and you support them."

— *Joshua Tarbutton, Founder and Chairman, Bravo Team*

CHAPTER SEVENTEEN:
Protection and Scale

By the time you reach this point, the architecture is designed. Authority has been placed upstream. The timing has been adjusted so that early action is legitimate. Learning mechanisms exist to capture reasoning before outcomes distort it. On paper, judgment now has somewhere to live.

What determines whether it survives is protection.

Every organization that claims to value judgment reveals what it truly believes when someone uses it for the first time under uncertainty. The response to that moment, more than any policy statement or leadership speech, teaches the organization whether early action is expected or merely tolerated. People do not listen to what leaders say about judgment. They watch what happens to the person who exercises it.

Most Smart Plants™ fail here. Early judgment is rhetorically welcomed and structurally exposed. Leaders encourage initiative. The system still rewards confirmation. The mismatch does not need to be announced. It becomes visible in the consequences.

Early action carries visible risk. Late action distributes responsibility across processes, alerts, thresholds, and escalation chains. That asymmetry shapes behavior faster than any training program ever will.

Why Early Judgment Feels Dangerous

Remember Marta hearing that bearing fail? That's early judgment. Acting before the dashboard turns red. Before the alert fires. When you can still prevent the problem instead of explaining it. Early action is personal. Late action is procedural. That asymmetry is why your people wait. They've learned that "too early" gets scrutinized. "Too late" gets excused.

Why did you pause the line? Why did you slow down the schedule? Why challenge the recommendation? Why now?

Those questions may be asked professionally and with curiosity, but they still create exposure. Late action rarely carries the same weight. When an alert fires or a threshold is crossed, the intervention appears procedural. The system required it. The process was followed. Responsibility diffuses.

Leaders often believe they are being fair when they evaluate early decisions with the benefit of hindsight. They review the outcome. They analyze the facts. They probe the reasoning. Yet fairness after the fact still introduces a quiet penalty. Information that was unavailable at the time of action now shapes the evaluation. The interruption becomes a reference point. The pause becomes a story retold. Over time, the organization absorbs a precise lesson: Early judgment must be defended. Waiting rarely must.

This is the trade most leaders never consciously name. Accountability remains intact. Initiative becomes costly. A costly initiative concentrates on a small group of people who can afford the friction.

If you want to know whether protection truly exists in your operation, look at the last three instances where someone acted early and turned out to be wrong. Not reckless. Wrong. What happened to them afterward? Were they evaluated on the reasoning available at the time or on the absence of a problem that never materialized? The answer to that question tells you more about your system than any engagement survey.

Protection Is Structural, Not Sentimental

Encouragement and psychological safety matter, but they do not override structural consequences. People do not decide whether to act early based on whether leadership seems supportive in principle. They decide based on pattern recognition: What usually happens to people who interrupt momentum here?

Protection becomes real only when evaluation criteria are explicit and consistent. Early action must be assessed against whether authority was exercised within its defined boundaries, whether the timing aligned with the operating logic, and whether the pause preserved optionality. Outcome alone cannot be the metric.

Once those criteria are stable, early judgment normalizes. It stops looking like heroism or disruption and starts looking like part of the work. Without that structure, the architecture you designed in Chapter Sixteen collapses under the first uncomfortable review.

The Organizational Math People Actually Do

Before acting early, people run a simple calculation. If I act now and I am wrong, what follows? If I wait and the system turns out to be wrong, what follows? In many advanced operations, the answer is to wait. Not because people lack commitment, but because the cost of visible initiative outweighs the cost of delayed intervention.

You cannot correct that imbalance with messaging. You correct it by changing what happens after someone exercises judgment.

Without protection, judgment becomes a luxury only your most bulletproof employees can afford. Everyone else? They're playing it safe because they can't afford to be right early and wrong on paper. And you're paying them to watch the trainwreck in slow motion. Signals arrive later. Concerns soften. Leadership does not lose intelligence entirely; it loses timing. And timing is the only thing that makes judgment valuable.

> **"How is your technology strategy impacting your HR strategy? If you want to hire and retain the best workers, you have to show them you're investing in technology to make workers' lives better, safer, easier - jobs using their brain and less using their brawn."**
>
> — *Will Healy III, Leader, Teradyne Robotics*

Scaling Judgment Without Slowing the Plant

Protection alone is not enough. Judgment must also distribute. Systems eliminate what does not scale, and judgment that lives inside a handful of experienced individuals does not scale. It compensates for design gaps quietly, sometimes brilliantly, but it remains concentrated rather than capable.

Walk your floor and identify the two or three people everyone consults when something feels off but cannot yet be proven. Ask yourself what specific interpretive load they carry. What patterns do they see? What timing instincts do they use? If those individuals left tomorrow, what capability would leave with them?

Most leaders can answer those questions intuitively. Few have translated the answers into design.

Judgment concentrates by default wherever authority is ambiguous, timing is late, and protection is inconsistent. Only those with insulation can afford to intervene. The rest learn to escalate or wait. Leadership often interprets this concentration of judgment as depth of expertise. In reality, it is dependency.

This creates a succession risk that no dashboard captures. When judgment lives in people rather than architecture, continuity becomes personal. Retirement, burnout, or attrition removes more interpretive capacity than any system outage. The system continues to run. Metrics remain stable. Months later, subtle problems begin surfacing that "never used to happen." What changed was not competence. It was the surgical removal of accumulated judgment that had been compensating for design gaps.

"Our education and workforce development director regularly calls additive the gateway to manufacturing. It's an easy space for people to get their head around, and everything about it is digital, which makes it easier to introduce to students since they've only known operating in that environment."

— John Wilczynski, Executive Director, America Makes

Tacit knowledge does not transfer automatically in automated environments. When work flows smoothly and interruptions are rare, there are fewer visible moments for interpretation to surface. Without deliberate capture and distribution, experience leaves without ever becoming infrastructure.

AI intensifies this pattern. As predictions improve, fewer people are needed to keep routine decisions moving. The remaining human judgment becomes higher-leverage and easier to concentrate on. The organization looks efficient. And it is – until novelty appears and the small group carrying interpretive weight is overloaded or absent.

The constraint is not the presence of judgment. It is its distribution.

Judgment scales when assigned to defined roles, bounded in scope, triggered by recognizable conditions, protected from outcome-based penalties, and captured as learning. When those elements align, early intervention resolves issues where signals first appear rather than funneling everything upward. Distribution reduces bottlenecks rather than creating them.

If you attempt to "push judgment down" without those conditions, distribution fails. People do not practice what the system does not require. And what they do not practice, they lose.

What Protection Actually Protects

Protection does not shield people from accountability. It shields the timing advantage judgment provides. It ensures that acting early does not become a career liability. It prevents architecture from degrading into theater the first time quarterly pressure tightens.

At this stage, you have built the framework. Authority has a location. Timing has a window. Learning has a mechanism. Protection determines whether it all survives real pressure. Because pressure will come. A quarter will tighten. Staffing will lean out. Someone will present a compelling case that a particular pause was unnecessary friction. Someone will argue that the system can handle it without human intervention. The logic will sound efficient and responsible.

That moment is the test.

If leadership yields reflexively, the organization learns that judgment is conditional. If leadership defends the architecture, even when uncomfortable, judgment becomes infrastructure.

The framework is built. The question now is whether it holds when the incentives push against it. Chapter Eighteen examines exactly that pressure, how judgment survives optimization, performance demands, and the constant internal argument that speed alone equals progress.

Judgment does not disappear in dramatic fashion. It erodes when leaders decide that this time, in this instance, the pause is not worth it. The final work of leadership is recognizing that those small decisions accumulate into architecture.

And architecture, once eroded, is far harder to rebuild than it was to design.

"There's the technology that runs our plant, and then there's the technology that runs our back office…and a lot of times they're still not communicating."

— *Dag Calafell, Director of Technology Innovation, MCA Connect*

CHAPTER EIGHTEEN:

When Systems Collide

At a bolt manufacturing facility I toured with the Industrial Fastener Institute, I watched a young operator struggle with a machine worth more than a million dollars. The equipment was sixty to seventy years old, rebuilt multiple times, still producing precision parts, and still profitable. The operator had been trained on the company's newer digital equipment, where adjustments are automatic and performance is displayed on touchscreens.

This machine had none of that. No digital display. No automated correction. Just dials, levers, sound, vibration, and decades of accumulated knowledge about how it actually ran.

"Young people can't run these," the tour guide told me. "They are trained on digital now. Nobody teaches manual anymore."

The company operated sixty pieces of equipment serving more than 1,400 active customers. Some were modern. Some were ancient. All were critical. They had just received approval for a new apprenticeship program the day before my visit, which was an acknowledgment that the gap between what the equipment required and what training provided had become a serious liability.

This was not a story about outdated machines. It was a story about systems colliding.

The organization had designed a capability for a single operating logic: digital, automated, and self-adjusting. It had allowed the other, older capability to wane: manual, interpretive, experience-driven. The two systems coexisted without coherence. One looked advanced. The other was fragile. Together they created a risk that was easy to miss because performance still held.

That is what a collision looks like in real operations. Strength in one dimension. Brittleness in another. Leadership often sees strength first.

Here's what most leaders miss: You can do everything right in isolation and still build something catastrophically fragile. Judgment is designed in one place, but authority lives in another. Timing that allows early action, but protection that punishes it. Learning systems that capture reasoning, but evaluation systems that care only about outcomes. Every piece works. Nothing aligns. And the system resolves that tension only the way it knows how: it chooses speed over sense. You built five good things that don't talk to each other, and you're calling it architecture.

Failure Rarely Comes from Bad Decisions

Smart Plants™ rarely fail because leaders make reckless choices. They fail because sensible decisions are made in isolation.

Judgment is required, but authority arrives too late to shape commitment. Authority is clarified, but timing still favors confirmation over early interpretation. Timing allows intervention, but protection remains inconsistent. Protection exists, but learning does not capture what early judgment revealed.

Each decision improves something. Together, they produce a system that moves quickly and adapts slowly.

You have spent the last several chapters designing the architecture. Judgment has a defined place. Authority has been moved upstream. Timing has been adjusted. Protection has been addressed. Learning mechanisms have been introduced. What this chapter forces you to confront is that alignment is not automatic. When these elements do not reinforce each other at the same decision points, the system resolves the tension on its own.

And it resolves it in favor of speed.

How Systems Resolve Misalignment

Organizations do not tolerate ambiguity in responsibility for long. When judgment, authority, timing, and protection are not aligned, the system reduces friction wherever possible. Interpretation gives way to escalation. Discretion gives way to proof. Discussion narrows into metrics. The operation stabilizes around what is defensible.

In that environment, people still notice. They still raise concerns. They still document observations. But action shifts elsewhere. Over time, judgment becomes compliance with process rather than interpretation of reality. Language softens. Concerns are framed cautiously. Silence increases, not because people disengage, but because they understand where authority truly sits.

From the outside, the plant appears mature. Interruptions decline, escalations decrease, and reporting looks disciplined. Inside, something important has changed. The ability to interrupt has been engineered out. Smoothness is not the problem; the inability to interrupt is. A system that cannot be interrupted early will eventually be interrupted later by something larger.

The Time-Horizon Failure Leaders Miss

Misalignment does not create immediate failure. In many cases, performance improves first. Variability drops. Escalations decrease. Leaders feel more in control.

The consequences surface later, when the system encounters something it was never designed to interpret. By then, the compensating behaviors are embedded. The people who once intervened early have learned that doing so carries a cost. Leadership experiences the outcome as sudden. It is not sudden. It is cumulative.

Here is the uncomfortable reflection. If your operation looks more stable than ever, but fewer people feel responsible for interrupting it, you may not be witnessing maturity. You may be witnessing compression. Compression feels like control. It often signals postponed learning.

AI Does Not Correct Misalignment

Artificial intelligence does not repair architectural misalignment. It amplifies it.

If judgment is required but authority arrives late, AI scales observation without intervention. If authority exists but the timing is still awaiting confirmation, AI narrows options more quickly. If timing allows early action but protection is inconsistent, AI accelerates silence. If learning fails, AI optimizes around incomplete understanding.

Technology will faithfully execute the structure you give it. If that structure sidelines human judgment at key commitment points, AI will make that sidelining more efficient. This is not an argument against automation. It is a reminder that automation magnifies coherence or incoherence. It does not fix either.

Succession Drift

When alignment is weak, judgment tends to concentrate in a small group of experienced people. Authority and interpretation collapse into informal roles. Leadership pipelines appear intact on paper, but interpretive capacity is carried rather than designed.

When those individuals leave, capability disappears abruptly. The system continues to run. Metrics remain steady. Months later, subtle problems begin surfacing that "never used to happen." Leadership assumes a discipline issue. In reality, accumulated judgment that once compensated for design gaps has walked out the door.

The loss is rarely recognized as architectural. It gets labeled as performance drift, cultural inconsistency, or a training gap. But what changed was structural. The organization was functioning on borrowed judgment, and when the lender left, the debt came due.

That is not a talent failure. It is a leadership design decision.

Why Partial Fixes Fail

Many redesign efforts stall because they address one element at a time. Encouraging judgment without moving authority creates frustration. Assigning authority without legitimizing early timing creates escalation. Adjusting timing without protection creates hesitation. Building learning systems without consistent evaluation creates caution.

Each fix appears reasonable. Each leaves another element exposed. The system compensates, and whatever had improved begins to erode again.

> **"Manufacturers are stuck in a quagmire of proprietary protocols and custom software development."**
>
> *— Mike Bowers, Chief Architect, FairCom*

Alignment requires coherence at the same commitment point. Judgment must be required when decisions become difficult to reverse. Authority must exist there. Timing must legitimize early action there. Protection must govern how action is evaluated there. Learning must capture what the intervention revealed there.

When those conditions reinforce each other, behavior shifts without speeches. When they do not, behavior stabilizes around defensibility.

The Real Cost of Early Action

Early action is not free. It interrupts the flow. It consumes attention. It introduces ambiguity into environments that prize certainty. Leaders who claim to value early intervention but refuse to acknowledge its cost will not sustain it. Early costs are small, visible, and reversible. Late costs are larger, distributed, and compounding. Early costs feel discretionary. Late costs feel inevitable. That asymmetry trains waiting.

Leadership must decide which discomfort it is willing to absorb. The discomfort of early uncertainty or the cost of late consequence. There is no option where no one is uncomfortable. If you are unwilling to absorb ambiguity early, the organization will later absorb the consequences.

Outcome-Based Evaluation Destroys Prevention

Early action evaluated by outcome will not survive. When prevention works, it looks unnecessary. When it doesn't, it looks foolish. In both cases, hindsight punishes early judgment. This is why organizations say they want early action while training people to wait. Judgment retreats not because people lack courage, but because the evaluation logic makes waiting rational.

Think about your last safety near-miss investigation. Did anyone ask, "What made this person notice early?" or did the conversation focus on whether the concern was ultimately justified? If it's the latter, you've just taught everyone watching that early intervention gets scrutinized while late response gets excused.

The Cost of Silence

Most operations measure the cost of disruption. Few measure the cost of silence. Silence hides delayed interpretation, compressed decision windows, and lost optionality. Silence accumulates until leadership is forced to respond at scale.

Early action makes the cost visible while it is still manageable. Late action hides costs until they compound. If you are tracking only the cost of interruptions and not the cost of silence, you are running an incomplete accounting system.

When Design Stops Colliding

Collision ends when judgment, authority, timing, protection, and learning reinforce each other at the same decision points. When that alignment exists, escalation slows. Early signals surface. Interpretation becomes legitimate work. Speed remains, but it can be interrupted.

Once you see this, the practical question emerges. Where do you begin?

You do not attempt to correct everything simultaneously. You identify one commitment point where misalignment is currently costing you capability. You align judgment, authority, timing, protection, and learning there. You observe what changes. You treat the result as proof, not theory. Then you extend the pattern.

Think about your last three operational surprises. Not the root cause. The moment when system certainty and lived reality diverged. That divergence marks the point at which architecture failed. That is where alignment begins.

Here's your assignment: Pick one of those surprises. Now trace it backward. Where did judgment arrive before certainty? Where did authority exist upstream? Where was timing early? Where was protection structural? Where did learning capture reasoning?

For each requirement that was missing, ask yourself: Why? Was it not designed? Was it designed but not defended? Or was it designed, defended, and then optimized away when pressure built?

That answer tells you whether your problem is architecture or endurance.

Systems do not fall apart all at once. They erode one concession at a time. One "not this time." One exception that feels justified. One pause that seems expendable in the moment.

Alignment is only the beginning. The real test is whether you will defend it when pressure builds, incentives tighten, and speed once again looks more responsible than judgment.

Leadership decides whether those moments accumulate into collision or into resilience.

"Don't spend time collecting and analyzing the data; spend time doing what the data's telling you to do."

— *Vince Sassano, President, Strategic Performance Company*

CHAPTER NINETEEN:
The Entry Point

The False Choice Between Transformation and Stability

By this stage, most leadership teams stall for a predictable reason. The work feels too large. When judgment, authority, timing, protection, and learning are all discussed together, it sounds like transformation. Leaders imagine system overhauls, culture resets, new governance structures, and disruption that their operation cannot afford. The perceived choice becomes binary: tolerate fragility or destabilize everything in pursuit of resilience.

That choice is false, and it keeps leaders trapped by turning disciplined governance into a fantasy of transformation.

Smart Plants™ do not regain judgment by fixing the entire system at once. They regain it by deciding where the system is no longer allowed to proceed without human interpretation. That shift moves the work from transformation to governance. From overhaul to precision. From culture change to design intent.

The real question is not how to change everything at once. It is where the system must no longer be allowed to decide alone.

Coherence requires alignment, but it does not require simultaneity across the entire enterprise. Judgment returns through focus, not scale. You do not repair architecture by repainting the whole structure. You reinforce a load-bearing point first.

Why Judgment Must Return Selectively

Judgment does not need to exist everywhere to change behavior everywhere. It needs to exist where consequence accumulates. Every operation contains moments when decisions harden, options narrow, and time, capital, safety margin, or flexibility quietly trade hands. These moments are often routine. That is precisely why they matter.

> **"You're either going to change or die. That's the cold hard reality of manufacturing; it is moving quickly and we have to move with it."**
>
> *— Roger Atkin, President, National Tooling and Machining Association (NTMA)*

Commitment points are leverage points. When judgment is redesigned at even one such point, behavior around it shifts. Signals surface earlier. Escalations shorten. Conversations move upstream. The organization becomes interruptible where it matters and no slower where it does not.

Broad empowerment initiatives fail because they ask for judgment everywhere without changing the system anywhere. Precision redesign works because it makes judgment unavoidable at the places that shape downstream reality. You are not asking people to use better judgment. You are designing the work so the system cannot advance without it. That distinction changes everything.

Why Stable Processes Are the Right Starting Point

Leadership instinctively looks for broken processes to fix. Broken things demand attention. Stable ones do not. In this work, that instinct misleads.

The most effective place to reintroduce judgment is not where performance is unstable, but where performance appears reliable. Stable processes often run smoothly because automation, thresholds, and procedures have replaced interpretation. Decisions advance without pauses. Metrics remain green. Exceptions decline. From the outside, it looks like maturity.

Sometimes it is maturity. Often, it is compression.

When nothing is on fire, leaders can observe how authority, timing, protection, and evaluation behave without the distortion of urgency. You can see whether early intervention is structurally possible or merely theoretical. That clarity disappears in crisis.

Here is the reflection that matters. If your smoothest process has not been meaningfully interrupted in months, ask yourself whether that smoothness reflects strength or untested brittleness. Stability without interruptibility is not resilience. It is delayed exposure.

What Changes at the First Commitment Point

When leadership selects one commitment point and redesigns it appropriately, four things change at once.

First, the system is no longer allowed to proceed automatically. A specific human role must interpret conditions before commitment occurs. Not escalate. Not comment. Decide.

Second, early intervention becomes legitimate rather than exceptional. The pause is designed into the work, not treated as friction.

Third, evaluation shifts from outcome to reasoning. The question becomes whether the interpretation was sound, given what was known at the time, not whether the feared consequence materialized.

Fourth, learning has somewhere to land. Early judgment produces information that the system previously ignored. That information becomes part of the design.

What does not change is equally important. Technology remains. Dashboards remain. Metrics remain. Workflow remains. The system can still move quickly. It simply cannot move blindly through that point.

If your production release process runs efficiently for 95% of cases, that continues. The change occurs in the 5% where signal drift, new suppliers, or pattern variation introduce ambiguity. In those cases, the system waits for interpretation. That single boundary restores timing advantage without destabilizing the plant.

This is not a transformation. It is disciplined governance.

Why This Does Not Create Chaos

Leaders often fear that legitimizing interruption in one place will unleash it everywhere. In practice, the opposite occurs.

Most professionals are conservative. They do not want unlimited discretion. They want a definition of when discretion is required and when it is not. Vague empowerment creates noise. Clear design creates discipline.

When judgment is required at a defined commitment point, the rest of the system continues to operate efficiently. Only that decision pauses, and only when interpretation is warranted. Over time, the quality of interruption improves.

People learn to distinguish between situations the system can manage and those it cannot.

Discipline does not come from restraint. It comes from design. And design is a leadership decision.

The First Signal the Organization Watches

People rarely respond first to announcements. They respond to consequences. The first time someone exercises early judgment under the new design, the organization pays close attention.

Is reasoning evaluated or outcome judged? Is the pause supported or resented? Does leadership appear early or only after impact?

One poorly handled moment can undo months of preparation. One well-handled moment can establish credibility faster than any rollout plan. That first intervention is not simply a decision event. It is a cultural inflection point.

Before the redesign begins, the evaluation criteria must already be explicit. The person exercising authority must know how their reasoning will be assessed. Leadership must be visible in defending interpretation over hindsight.

Trust does not accumulate through repetition. It accelerates through evidence. When early action is handled correctly once, the message spreads laterally without instruction. People recalibrate quickly when they see that acting early does not create personal exposure.

Why Premature Scaling Fails

Once the first redesign appears to work, the instinct to scale is strong. Leaders codify, standardize, expand the scope, and formalize training. Done too quickly, this undermines the very capability just created.

Judgment infrastructure matures through use, not documentation. Authority, timing, and protection must be observed under pressure long enough to reveal where they wobble. Scaling before learning stabilizes and converts nuance into procedure. Rigidity returns, only with better formatting.

The pattern is familiar. The pilot works. Expansion accelerates. The new locations imitate the form without internalizing the discipline. Within months, behavior drifts back toward confirmation and escalation.

If you cannot articulate what made the first commitment point work beyond surface mechanics, you are not ready to replicate it. You are copying the structure without understanding the load.

Readiness Is Structural, Not Emotional

Readiness is not enthusiasm. It is not an agreement. It is not urgency. Readiness is the system's ability to absorb early ambiguity without penalizing it.

Organizations fail here not because the concept is unclear, but because the system cannot yet sustain the tension that early judgment introduces. When early action yields no obvious outcome, leadership must decide whether to defend the reasoning or question its necessity. That moment reveals readiness.

If you cannot describe how someone will be evaluated when they act early and are later proven wrong, you are not ready. If you cannot name the specific role holding authority at the chosen commitment point, you are not ready. If leadership is not prepared to publicly defend its reasoning when the outcome does not justify it, you are not ready.

Launching a redesign in those conditions does not build resilience. It trains caution. Better to prepare the ground than to introduce judgment into soil that will reject it.

Capability and Capacity

Most organizations already possess judgment capability. Experienced operators understand patterns. Supervisors know when something feels off. Leaders can decide without full information. What they lack is capacity.

Capability is the ability to interpret. Capacity is the system's ability to hold interpretation repeatedly without exhausting trust, attention, or political capital. Capability without capacity produces heroes. Capacity without capability produces bureaucracy. Resilience requires both.

The entry point exists to test whether capacity can be built alongside capability. Without that test, judgment remains dependent on individual courage rather than design.

The First Ninety Days

This is not a transformation plan. It is a contained test of whether judgment can survive contact with your system.

Select one commitment point with a meaningful but manageable consequence. Ensure the boundary between before-and-after commitment is identifiable. Confirm that multiple weak signals exist, not merely a single threshold. Identify a specific role that can plausibly hold authority without escalation.

Before redesigning, document reality. Who decides today? When does commitment occur? What triggers action, and what is ignored? How are early concerns evaluated, if at all? Where does learning accumulate or disappear?

Only after reality is visible should redesign begin. Make four conditions explicit at that single point: the system may not proceed without human interpretation; a defined role holds decision authority; early intervention is legitimate within a stated window; and evaluation will focus on reasoning rather than outcome.

Implement quietly with one team. Leadership presence must be visible. Protection must be explicit before the first decision. If outcome-based evaluation reappears, correct it immediately.

For the next several weeks, prioritize observation over efficiency. Where did judgment operate as designed? Where did the system resist? Where did protection weaken under pressure? What surprised you about timing or authority?

If friction appears earlier and executive surprise decreases, the design is functioning. Discomfort at this stage is not regression. It is the sensation of interruptibility returning.

Do not expand until you can answer this clearly: What specifically made judgment work here, and what would cause it to erode if replicated carelessly?

If that answer is vague, stop. You're not ready to scale.

And if you're thinking, *We don't have time to be this deliberate,* then be honest about what you're really saying: *We'd rather move fast and discover our fragility under pressure than move deliberately and prevent it.*

That's a choice. Own it.

Why This Is the Only Sustainable Entry Point

Smart Plants™ do not mandate judgment back into existence. They restore it through precision. They choose a single commitment point where the system is no longer allowed to advance without human interpretation, and they align authority, timing, protection, and learning at that boundary. One intervention handled correctly, one visible defense of reasoning over hindsight, and one protected decision evaluated on logic rather than outcome can shift the organization's understanding of what is truly valued.

From that point forward, judgment does not need to be forced across the enterprise. It spreads because the system has demonstrated that interpretation has a legitimate place to operate. When people see that early action does not create personal exposure, they adjust. When leaders consistently defend reasoning over results, credibility compounds. Design becomes intentional again – not through broad declarations, but through disciplined containment and visible reinforcement at the moments that matter.

The framework is now complete. The entry point is identifiable. The conditions are knowable. What remains is not architecture but resolve.

Alignment restores judgment. Endurance determines whether it lasts.

Restoring judgment at one commitment point is the beginning, not the finish. The real work is sustaining it when incentives tighten, performance pressure increases, and speed once again becomes the more attractive option. Leadership will be repeatedly asked whether the pause is still worth it. The answer to that question determines whether judgment becomes infrastructure or fades back into exception.

"We have to be quick, adaptive, and responsive. Yet at the same time, we need to be reflective. Stop and think...reassess the landscape because it's changed."

— *Craig James, Managing Partner and Co-Founder, Cat-Strat Services*

CHAPTER TWENTY:

When Success Becomes the Threat

Failure doesn't silence judgment. Your confidence does. When your systems perform consistently, you relax. Variance narrows, exceptions decline, and you start to believe the operation is self-sustaining. It's not. What you're watching is the difference between smooth execution and actual resilience. And you won't know which one you built until something breaks in a way your system never anticipated.

Here's what happens next: Judgment doesn't disappear. It loses its standing. The conditions that once made it legitimate – the visible risk, the recent failures, the urgency – fade into memory. Nothing appears at risk. So why would anyone interrupt?

When Success Rewrites Risk

Success alters perception long before it alters structure. What once required active protection begins to look contained. What once demanded scrutiny begins to look routine. Quiet carries authority.

When nothing breaks, judgment must justify itself. Dashboards stay green. Interventions resemble friction instead of care. Discretion begins to look indulgent. Risk does not vanish. It simply becomes invisible to anything not structurally required to notice it.

At one of the largest limestone quarries in North America, operating continuously since the 1840s, the performance indicators were exceptional. Turnover hovered between 1% and 2%. Rotary kilns ran continuously. Millions of tons moved annually. By traditional measures, the operation looked stable and mature.

Beneath that stability, however, 35-40% of the workforce had more than thirty years of tenure. The operation was not failing. It was transitioning. And the transition was unfolding while every visible metric remained strong.

Retention, once a marker of resilience, had become a demographic cliff. Deep institutional knowledge was aging simultaneously. Succession was not spreading gradually. It was compressing. Nothing had broken. That was precisely the problem.

"We're the most visible invisible industry. You complain if a sign isn't right, but you take for granted when it works."

— Lori Anderson, President & CEO, International Sign Association (ISA)

The plant had not become fragile because performance slipped. It had become fragile because performance concealed structural drift. Success hid the erosion of distribution, transfer, and overlap. The conditions that supported resilience were narrowing beneath flawless execution.

That is the danger of sustained success. It rewrites the story of risk.

The Illusion of Resilience

Organizations frequently confuse repetition with resilience. If the system performs reliably, leadership assumes it can adapt reliably. But reliability reflects execution under known conditions. Adaptability governs response when conditions shift.

Success strengthens execution. It does not automatically strengthen adaptation.

Without governance, nothing forces leadership to maintain that distinction. Judgment exists to hold that boundary in place. When judgment is not required, the distinction collapses into assumption. And the first time the system encounters novelty, leadership discovers that repetition was mistaken for range.

Here is the uncomfortable question: Are you measuring how well your plant repeats or how well it adapts?

If your metrics overwhelmingly reward consistency, throughput, and utilization, you are measuring repetition. That is not wrong. It is incomplete.

How Flow and Proof Displace Judgment

As systems mature, decisions migrate into defaults. Escalation paths tighten. Intervention thresholds rise. Decision rights consolidate. These shifts are framed as improving flow. They often do. Defaults win because they require no permission.

Over time, the organization stops asking where judgment is required and assumes the answer has already been settled. The deliberate pause becomes an anomaly. Explanation becomes a burden. In successful systems, the need to justify interruption starts to feel like inefficiency.

What disappears first is not judgment itself, but permission to use it.

People still notice anomalies. They stop acting on them because the system advances without their interpretation. Flow becomes the dominant value. Flow is measurable. Flow is defensible. Flow is rewarded.

Your best people understand this faster than anyone else. They see what the system legitimizes. They adjust. If the system rewards continuation, they stop pausing. The capability you will need during disruption is trained out during stability. That is not disengagement. That is alignment.

When Optimization Becomes Authority

Optimization rarely looks threatening. It arrives as responsibility. It promises efficiency, consistency, and measurable improvement. In mature systems, it becomes indistinguishable from good management. That is exactly why it is so powerful.

Optimization does not simply improve outcomes. It reshapes legitimacy. It determines what the system treats as real, valuable, and actionable. Judgment interrupts. It hesitates. It reframes when signals conflict. It operates before certainty exists. Optimization acts after confirmation. It rewards what can be measured, repeated, and scaled. This creates an asymmetry. Judgment prevents. Prevention looks like nothing happened.

When someone asks, "What did that pause accomplish?" and the honest answer is, "We avoided a problem that never materialized," optimization hears cost. Time was consumed. Resources were spent. The avoided consequence is invisible precisely because it did not occur.

Over time, improvement itself becomes authority. If a decision increases throughput or utilization, it carries automatic legitimacy. If a decision preserves interpretive capacity but does not move a metric, it must defend itself.

Consider your last leadership meeting. How many decisions were justified because they improved a number? How many were justified because they preserved judgment capacity? If the ratio is heavily skewed toward improvement, optimization has already become your authority structure. Judgment is not debated in that environment. It is routed around.

When Erosion Becomes Structural

Every improvement cycle narrows decision space slightly. Slack disappears. Decision windows shorten. Optionality compresses. Early in a system's life, this feels productive. Noise declines. Priorities sharpen. Over time, the space between "The system says proceed" and "We have committed resources" shrinks. That space is where judgment lives.

Optimization is context-agnostic. Judgment is context-dependent. Standardization drives efficiency at scale. It also reduces the influence of local interpretation. Without explicit governance, context is treated as variance to be eliminated rather than as information to be weighed.

Many leaders believe culture can balance this tension. It cannot. Optimization has structure, metrics, and momentum behind it. Judgment without formal protection is simply intention competing with process. As scale increases, authority relocates from interpretation to system flow. Humans remain involved, but later. By the time judgment engages, optimization has already shaped the path.

> **"Manufacturing can't afford downtime. If the cloud goes down, you stop processes dead."**
>
> — *Mike Bowers, Chief Architect, FairCom*

Reintroducing pause in that environment feels like regression. Restoring discretion feels like inefficiency. The system resists mechanically because it was designed to preserve gains. Scale does not create erosion. It locks it in.

The Governing Constraint

Optimization cannot determine which risks are unacceptable. It cannot weigh novelty against historical performance. It cannot recognize when long-term success has created structural vulnerability in silence. Those decisions belong to leadership, and they must be made when nothing appears wrong, because that is when erosion accelerates most aggressively. If you wait for evidence of decline before governing optimization, you are already late.

So here are the decisions that remain: What will you refuse to optimize? Where will you declare that judgment is non-negotiable, even when metrics improve? Where will you absorb the discomfort of early interruption rather than defer the cost to a later correction?

Every organization pays for judgment. The only variable is timing. Either leadership governs optimization, or optimization governs leadership. And here's what that payment looks like when you choose to pay late.

What Erosion Actually Costs

You've read twenty chapters about how judgment erodes. You understand the mechanics. You can see it happening in your operation. But if you're still thinking of this as a cultural issue or a leadership development challenge, you're missing what it costs. Not in engagement scores or retention statistics. In money. In margin. In the kind of exposure that shows up in quarterly calls and insurance audits.

Judgment erosion doesn't announce itself as a line item. It hides inside "one-time" expenses, unexpected remediation costs, and operational surprises that leadership explains as isolated incidents. They're not isolated. They're compounding. And they're expensive in ways most executives don't connect until the pattern becomes undeniable.

Here's what judgment erosion costs when you stop measuring culture and start measuring cash.

The Exponential Remediation Multiplier

When judgment operates early, problems are caught while they're still cheap to fix. A bearing sounds off. Someone pauses the line. Maintenance investigates. A $1,200 part gets replaced during a scheduled window. Total cost: $1,200 in parts, two hours of labor, zero production loss.

When judgment has eroded, that same problem progresses undetected. The bearing fails catastrophically. Debris moves through the system. Secondary damage occurs before anyone realizes the scope. The line shuts down for six days. Now you're looking at $340,000 in repairs, $180,000 in expedited shipping, $95,000 in customer penalties, and $420,000 in lost production.

A $1,200 problem just became a $1,035,000 problem. That's an 862x cost multiplier.

That's not an exaggeration. That's Oakridge. And Oakridge is not fictional in its cost structure; it's a composite of actual failures I've documented across dozens of facilities where early signals existed, were noticed by someone, and went nowhere because the system didn't legitimize action before proof.

The remediation multiplier grows exponentially the later judgment enters the sequence:

- Early detection (judgment active): 1x base cost
- Threshold detection (system catches it): 10-25x base cost
- Failure detection (damage occurs before response): 100-500x base cost
- Cascade detection (multiple systems fail before containment): 500-2000x base cost

Every day you operate without judgment infrastructure, you're raising the stakes on the next surprise. You're not avoiding cost. You're deferring it with interest.

Customer Confidence Erosion

Customers don't care about your internal metrics. They care about one thing: Can you be trusted to deliver what you promised, when you promised it, at the quality you committed to? A single "surprise" failure resets that trust to zero.

Not a flagged issue that you managed proactively. Not a forecasted delay that you communicated early. A surprise, the kind where the customer finds out something went wrong because their order didn't show up, or worse, because their downstream process failed using your defective component.

That failure doesn't just cost you the immediate order. It costs you:

Immediate revenue loss: The cancelled or returned order, the expedited replacement costs, and the penalty clauses in your contract that activate when delivery or quality commitments are missed.

Qualification costs: Two of Oakridge's major customers added backup supplier requirements after the Line 4 failure. That means Oakridge now competes on price alone for 30% of the volume they used to own on relationship and performance. Over eighteen months, one customer shifted that 30% entirely to the backup supplier. Revenue impact: $4.2M annually.

Recertification costs: In regulated industries, a single quality failure can trigger recertification requirements. You're now paying for third-party audits, process documentation reviews, and sometimes facility upgrades to prove you've addressed root cause. Budget: $150,000 to $500,000, depending on industry and scope.

Margin compression: Even customers who stay with you will renegotiate. They'll demand price concessions, tighter SLAs, or enhanced reporting in exchange for continuing the relationship. That margin compression persists for years, long after you've "fixed" the problem.

Add it up, and a single judgment failure that creates a customer surprise can cost you $5M to $15M in direct and opportunity costs over a three-year period.

And that assumes you don't lose the customer entirely.

If you lose a major account because trust collapsed after a preventable failure, calculate what replacing that revenue actually costs: sales pursuit expenses, qualification cycles, onboarding complexity, and the reality that new customers almost always negotiate harder than customers you have served for years.

Recruiting and Reputation Impact

Manufacturing is a small world. When your plant has a major failure, people hear about it. Not just customers, but also potential employees.

The Oakridge Line 4 failure made regional news. It wasn't a safety incident. It wasn't a recall. It was a six-day unplanned shutdown at a facility that had just celebrated 500 consecutive days of uptime. That story spread.

Six months later, Oakridge's recruiting pipeline had dried up. Experienced operators at competitor facilities who had previously considered Oakridge a stable, desirable employer were no longer interested. The narrative had shifted from "Oakridge runs a tight operation" to "Oakridge looked good until something broke; they should have seen it coming."

That shift is expensive.

Increased recruitment costs: When your reputation shifts from "stable" to "brittle," you have to work harder to attract the same quality of talent. Job postings stay open longer. Offer packages increase. Signing bonuses become necessary. Recruiting agency fees rise because placements are now harder to secure. Cost increase: 35-50% per hire.

Quality of hire declines: The best people have options. When your facility's reputation takes a hit, you're no longer competing for top talent; you're competing for who's left. That means longer onboarding, more turnover in the first eighteen months, and higher training costs per productive employee.

Turnover acceleration among existing staff: When a major failure occurs, your best people start updating their résumés. Not because they're disloyal, but because they recognize that a facility unable to prevent obvious failures has leadership problems. Turnover among high performers costs you institutional knowledge, increases training burden on those who remain, and creates operational gaps that further erode judgment capacity.

Oakridge lost four senior operators in the eight months following the Line 4 failure. Combined tenure: 127 years. Replacement cost, including lost productivity during transition: $890,000.

And here's the part most leaders miss: reputation damage doesn't heal quickly. It takes years of flawless execution to rebuild a reputation for stability.

During that rebuilding period, you're operating at a recruiting and retention disadvantage compared to competitors who didn't experience a visible failure.

Insurance and Liability Exposure

When judgment erodes, risk concentrates. Not gradually, but suddenly, during the next event the system wasn't designed to handle. And when that event occurs, your insurance carrier pays attention.

Premium increases: A single significant incident triggers an underwriting review. If the failure analysis reveals that early warning signs existed but weren't acted on, you're now considered a higher-risk organization. Premium increases of 15-40% are common after a major operational failure. On a $2M annual insurance spend, that's $300,000 to $800,000 in increased costs annually.

Coverage restrictions: Insurers don't just raise rates; they narrow coverage. Exclusions get added. Deductibles increase. Co-insurance requirements tighten. You're now carrying more risk on your own balance sheet because your insurer no longer trusts your operational controls.

Regulatory exposure: In some industries, failures stemming from ignored early warnings can trigger regulatory scrutiny. OSHA investigations. EPA reviews. Industry-specific compliance audits. Even if you're not fined, the cost of responding to a regulatory inquiry – legal fees, documentation production, and executive time – runs into six figures quickly.

Litigation risk: If your judgment failure created a downstream impact on a customer's operations, you're now exposed to breach of contract claims, consequential damages, and in extreme cases, negligence arguments. Legal defense costs alone can exceed $500,000 before you ever reach a settlement or judgment.

The compounding effect is what matters. A single failure rooted in judgment erosion doesn't just cost you once. It recalibrates your risk profile in ways that increase costs for years.

Succession Costs When Experience Walks Out

You have three people in your operation right now who carry disproportionate interpretive load. Everyone knows who they are. They're the ones people go to when something doesn't make sense. When the data conflicts with reality. When a decision needs to be made without complete information.

What happens when they leave?

Most leaders assume knowledge transfer plans will handle it. Documentation. Cross-training. Overlap periods. In practice, those mechanisms capture outcomes. They don't transfer judgment.

Hiring replacement costs: Finding someone with equivalent experience in today's market isn't just expensive, it's often impossible. Even if you find someone with similar tenure, they don't carry your organization's context. The patterns they've learned elsewhere don't map cleanly to your equipment, your customers, your processes. Replacement cost for a senior operator or technician: $75,000 to $150,000 in recruiting, onboarding, and lost productivity during ramp-up.

Institutional knowledge loss: When experienced employees leave, they take with them the memory of what almost went wrong. The fixes that aren't documented. The workarounds that prevent problems the system doesn't track. That knowledge isn't recoverable. Once it's gone, you don't know what you've lost until the next time you need it, and it's not there.

Increased failure frequency: After key individuals depart, failure rates often increase by 20-35% in the areas they used to oversee. Not because the replacements are incompetent, but because they haven't yet developed the pattern recognition that prevents issues from escalating. Each incremental failure carries cost: downtime, rework, scrap, and delay.

Leadership distraction: When experienced judgment leaves, problems that used to be handled at the operational level begin to escalate to leadership. Decisions that used to be resolved quickly now require meetings, analysis, and executive attention. The opportunity cost of senior leadership time being consumed by operational issues that used to be managed independently is high and rarely quantified.

Oakridge lost Frank Delgado in September 2028. Retirement. Well-earned. Thirty-four years of service. Leadership redistributed his responsibilities across three people and assumed continuity. By mid-2029, maintenance costs had increased by 18% and unplanned downtime had doubled in the systems Frank oversaw. Total cost of that transition: $1.4M in the first year alone.

Multiply that across the succession events coming in the next five years, because they are coming, and you're looking at a multi-million-dollar exposure that most facilities are not planning for, budgeting for, or even measuring.

The Bottom Line

Judgment erosion is not a soft issue. It's a financial issue that shows up in:

- Remediation costs of 100x to 2000x higher than early intervention
- Customer revenue losses of $5M to $15M per major trust failure
- Recruiting cost increases of 35-50% and quality-of-hire declines
- Insurance premium increases of 15-40% annually after incidents
- Succession costs exceeding $1M per critical departure

Add those together, and a single judgment infrastructure failure can easily cost a mid-sized manufacturer $10M to $25M in direct, opportunity, and compounding costs over a three-year period.

The question is not whether you can afford to protect judgment. The question is whether you can afford not to. Because the next time something breaks in a way your system didn't anticipate, you're going to pay. The only variable is whether you pay early in design and protection costs, or late in remediation and recovery costs. One is expensive. The other is catastrophic.

By this point, the framework is complete. You understand how judgment erodes, how to design it back in, how to protect it, and how to scale it. You've seen what it costs when you don't.

What remains is endurance. Not in the sense of surviving hardship, but in the sense of sustaining discipline when nothing appears broken. Because resilience is not built when systems are struggling. It is preserved when systems are thriving.

And preservation, as you're about to discover, is more difficult than design.

"If you don't constantly challenge the organization... there's an end of that runway that comes pretty quickly."

— *Kelly Springer, President & CEO, Metal Flow Corporation*

CHAPTER TWENTY-ONE:
Governance Fatigue

Success itself becomes destabilizing. Optimization narrows variance. Systems improve. Friction declines. What once required judgment now operates on momentum. The danger at that stage is no longer visible breakdown. It is erosion.

The next stage in that erosion is governance fatigue.

Governance does not usually collapse under pressure. Pressure sharpens focus. It forces leaders to pay attention, to intervene, to reassert authority. The more common pattern is the opposite. Governance weakens when performance is stable and results are strong. When operations are smooth, escalation becomes rare. Exceptions decline. Interventions feel unnecessary. Oversight continues formally, but the insistence behind it softens.

This shift does not happen because leaders stop caring. It happens because stability changes perception. When nothing appears broken, the effort required to maintain constraint begins to feel disproportionate to the visible risk. Meetings still occur. Dashboards are still reviewed. Authority structures remain in place. What changes is the willingness to use them to interrupt forward motion when the data does not demand it.

Why Fatigue Is Structural

Governance fatigue is not primarily a character issue. It is structural. Sustained success reduces visible consequences. Reduced consequence weakens feedback. When feedback weakens, the perceived necessity of active constraint declines. Over time, governance activity begins to feel ceremonial rather than essential.

In the early stages of system design, friction is obvious. Escalation paths are exercised. Intervention is frequent. Leaders are directly involved because risk is tangible. As performance improves, those friction points diminish. The absence of disruption is interpreted as proof that governance can operate more lightly. What is overlooked is that governance was never designed only for visible stress. It exists to protect judgment before stress becomes measurable.

Unless governance is deliberately designed to survive stability, oversight drifts toward confirmation rather than constraint. Leaders confirm what the system reports rather than challenging where it may be blind. That transition is gradual and rarely acknowledged.

Walk your floor and consider where someone has unquestioned authority to slow work when dashboards are green. Not where policy says they can, but where it happens in practice. If that authority feels theoretical, governance has already begun to fatigue.

When Monitoring Expands

One of the clearest signs of fatigue is the expansion of monitoring.

As systems scale, dashboards become more sophisticated. Reporting becomes more granular. Leaders gain deeper visibility into operational metrics. This appears to strengthen oversight. In reality, monitoring and governance serve different functions. Monitoring observes outcomes. Governance constrains action before outcomes are fixed.

Monitoring scales easily. It does not require confrontation. Governance requires authority that can interrupt momentum. It requires the willingness to slow production when the evidence is incomplete. When oversight drifts toward observation rather than constraint, performance can continue improving while the system quietly loses the capacity to correct itself early.

Leaders often feel more engaged during this phase because they are reviewing more data than ever before. Information, however, does not preserve authority. Being informed about system performance is not the same as governing the boundaries within which that performance occurs.

The relevant question is not how many dashboards are reviewed, but when leadership last intervened to protect judgment before measurable deviation occurred. If those moments are rare, monitoring has replaced governance.

The Backward Drift of Oversight

Fatigued governance shifts its timing.

Instead of intervening upstream, oversight becomes retrospective. Reviews focus on what happened. Lessons are drawn from confirmed outcomes. Intervention is justified by data that has already materialized. This posture feels disciplined because it is evidence-based. The organization appears thoughtful and analytical.

The difficulty is that judgment operates before certainty exists. It interrupts when confidence is still high. When governance activates only after confirmation, it reacts rather than protects. The timing of the intervention has moved downstream.

In stable systems, this shift feels efficient. It avoids unnecessary disruption. It preserves throughput. But it also narrows the window in which judgment can meaningfully alter direction. If intervention requires proof, proof will arrive after momentum has committed the organization to a path.

This backward drift is subtle because nothing looks neglected. Oversight still exists. The question is where it sits in time.

Stability as Permission

Sustained stability creates implicit permission to simplify governance.

Escalation pathways narrow because they are rarely used. Exception-handling processes atrophy. Intervention becomes unfamiliar. Authority that exists but is seldom exercised weakens in practice, even if it remains intact on paper. Over time, people hesitate to slow production because doing so now requires stronger justification than it once did.

This degradation will not appear in policy documents. It surfaces in small moments. An operator senses something unusual but waits because the metrics remain within tolerance. A supervisor chooses not to escalate because the system shows no alert. A maintenance concern is deferred because no threshold has been crossed.

Authority that is not exercised eventually feels unavailable.

Look closely at your incentive structure. Who absorbs the visible cost of slowing work? Who is rewarded for maintaining an uninterrupted flow? If continuity is consistently rewarded and interruption is consistently penalized, governance will yield to throughput under stable conditions. Even well-intentioned leaders will hesitate to intervene when success appears to justify speed.

Delegation and Diffusion

As fatigue progresses, responsibility diffuses.

Senior leaders expect managers to handle issues locally. Managers expect systems to surface meaningful exceptions. Systems detect what they are designed to detect. Judgment falls between these layers. Governance cannot be effectively delegated to a level at which authority to interrupt has already been transferred elsewhere.

Responsibility without authority produces compliance. It does not produce protection.

Incentives reinforce this diffusion. When performance metrics are strong, leaders focus on strategic initiatives. Managers manage daily execution. Systems manage micro-adjustments. No single layer feels responsible for challenging the system's direction when everything appears stable. Governance becomes assumed rather than tested.

If no one is explicitly accountable for verifying that governance still holds during periods of success, drift becomes structural.

The Compression of Time

Success also compresses decision timing. Actions that once warranted pause are resolved quickly because delay feels inefficient in a high-performing system. Speed becomes associated with competence. Leaders pride themselves on decisiveness.

Under these conditions, governance that introduces friction is tolerated primarily when failure appears imminent. When the system is performing well, intervention feels intrusive. If pause is legitimate only in a crisis, governance has become conditional rather than continuous.

Over time, the organization loses the habit of interrupting early. Intervention becomes a reaction to stress rather than a disciplined act of protection.

The Structural Trap

Governance becomes hardest to sustain at the moment it appears least necessary. Success removes urgency. Scale diffuses ownership. Optimization rewards continuity. Fatigue follows.

This pattern does not signal incompetence. It reflects how human attention responds to stability. Without governance deliberately constructed to resist fatigue, judgment erodes even in disciplined organizations. The smoother systems operate, the more intervention feels like interference. Leaders step back slightly.

Optimization fills the space. Friction declines.

If judgment depends on exceptional leaders stepping in heroically during a crisis, it has not been institutionalized.

What Must Endure

Judgment survives only where governance does not depend on vigilance alone. It endures where authority is embedded in operating architecture, where intervention is defined rather than discretionary, and where oversight persists even when attention shifts elsewhere.

Return to the commitment points you identified earlier. Are they still functioning as originally structured, or have they drifted toward speed and continuity? Who is responsible for testing governance integrity during stable periods? If that responsibility cannot be clearly named, governance has already softened.

Here's what makes this permanent: Someone owns it. Not a committee. Not "leadership." One person whose job includes verifying that the commitment points you protected are still protected. One person who asks every quarter: "Where did we let this slip?" One person who can say no when someone argues the pause isn't necessary anymore.

If you can't name that person right now, your governance is already periodic. And periodic governance is just scheduled erosion with better optics.

No organization sustains urgency indefinitely. Systems outlast attention. Governance must therefore be built to withstand routine, confidence, and distance from crisis. Design alone is not enough. It must become durable enough to survive fatigue, turnover, expansion, and succession.

Optimization makes judgment vulnerable. Stability makes governance vulnerable. The remaining challenge is permanence.

"The people that have been in those roles for twenty or thirty years have that legacy knowledge of how the company got started, why we do what we do; all of that walks out the door when that baby boomer finally decides to retire and takes all that knowledge with them."

— *Meaghan Ziemba, Host, Mavens of Manufacturing Podcast*

CHAPTER TWENTY-TWO:
The Three Losses

Governance fatigues under success. Even disciplined leaders cannot rely on vigilance alone to preserve judgment. Structure must endure when urgency fades.

There is a further reality, however. Even when governance is deliberately designed and actively maintained, judgment does not automatically sustain itself. Three permanent conditions work against it: succession resets legitimacy, memory displaces experience, and AI accelerates forgetting.

These are not episodic disruptions. They are structural forces present in every modern operation. Organizations that fail to account for them discover, often too late, that judgment was never as durable as it appeared.

Succession Is Not Continuity

Succession rarely registers as a risk event. It presents as continuity. Titles transfer. Roles are filled. Meetings proceed. Production continues. From the outside, the system appears unchanged. That surface stability is precisely what makes succession one of the most reliable failure modes for judgment.

Judgment does not transfer automatically with authority. It does not persist through org charts, documented processes, or strategic intent. When leadership changes hands, legitimacy resets. What appears to be seamless operational continuity often conceals a deeper discontinuity beneath. Formal authority transfers on day one. Judgment legitimacy does not.

Judgment operates because the organization recognizes who can interrupt, when interruption is warranted, and what consequences follow. That recognition is built over time through repeated action. It is anchored in memory, reinforced by outcomes, and stabilized by relationships. Succession disrupts all three at once.

New leaders inherit authority immediately. They do not inherit standing at the same pace. The system responds cautiously. People wait. Intervention slows. Discretion narrows. Concerns are packaged more carefully. Early signals are filtered rather than surfaced directly.

This recalibration is not resistance. It is risk management.

Consider the last time leadership changed in your operation. How long did it take for people to raise concerns with the same directness they had used previously? If it took months, that gap represented judgment operating conservatively. During that period, early intervention was possible but less likely.

Governance fatigue compounds this problem. If governance has already softened under success, succession accelerates the erosion. Without explicit reassertion of legitimacy, conservatism becomes the new baseline.

The Loss of Tacit Authority

Much of what sustains judgment is tacit.

It lives in unspoken agreements about timing, tolerance, and trust. It depends on accumulated credibility. It is reinforced by memory of prior interventions and their consequences. People act early because they know early action will be protected.

Tacit authority does not survive turnover intact.

When leaders leave, they take not only their decisions but also the credibility that enabled judgment to operate upstream. Successors must rebuild that standing. Until they do, judgment operates cautiously. Without governance that explicitly restores permission to interrupt, conservatism hardens into default behavior.

Organizations often attempt to preserve continuity through documentation. Processes are codified. Decision frameworks are archived. Lessons learned are recorded. These artifacts capture outcomes. They do not preserve legitimacy.

Judgment depends on permission to act before certainty exists. Documentation legitimizes action after the fact. It cannot confer authority in advance.

You can hand your successor a comprehensive decision manual. If you do not explicitly transfer legitimacy to interrupt the system when metrics are green, they inherit the structure without the standing.

Systems preserve flow. They do not preserve legitimacy.

Memory Displaces Experience

Organizations assume that what has been learned will remain available. Decisions are recorded. Data is stored. Systems retain history. From the outside, memory appears durable.

Judgment does not depend on information alone. It depends on lived consequence. It forms through proximity to uncertainty, through discomfort, through miscalculation and recovery. Experience embeds interpretation in a way documentation never can.

As experienced people leave, information remains, but consequences fade. The organization remembers what happened but gradually forgets why it mattered. Signals that once demanded interpretation become data points detached from context. Memory without consequence does not preserve judgment. It replaces it.

This replacement feels efficient. Lessons become procedures. Thresholds are formalized. Escalation rules are defined. Feedback loops shorten. What was once interpretive becomes procedural. Efficiency improves. Variability declines. What disappears is the friction that once required judgment to operate.

When learning is compressed into automation, judgment is no longer exercised. It is referenced. Over time, reference replaces experience. An organization operating on reference instead of lived interpretation is one unfamiliar event away from discovering that its people no longer know how to think independently of the system.

Governance fatigue makes this thinning harder to detect. If leaders are already relying more heavily on monitoring than intervention, the shift from experience to abstraction passes quietly.

AI as an Accelerator

AI does not introduce memory loss. It accelerates it.

AI systems learn from patterns that can be encoded. They reinforce what has worked and suppress what cannot be proven. They remember success exceptionally well. They do not remember hesitation, unease, or near-misses unless those moments are deliberately preserved.

Organizations often assume that once knowledge is encoded, it is secured. Models are trained. Decision trees are refined. Playbooks are updated. Yet encoding outcomes does not encode the legitimacy to interrupt. Storing data does not grant the right to hesitate.

As experienced employees leave, AI efficiently fills gaps. Decisions appear consistent. Performance holds. The organization interprets this stability as proof that capability remains intact. What disappears is the memory of fragility.

Without people who remember when judgment mattered and why, the system loses sensitivity to early signals. AI-driven systems learn by confirming themselves. Outputs reinforce inputs. The longer a system operates without visible failure, the more confident it becomes in its own logic.

Judgment once disrupted that loop. It questioned assumptions before evidence accumulated. It slowed decisions based on unease, not proof. When the legitimacy to question weakens, the loop closes. Closed loops feel stable. They are brittle.

Metrics do not reveal this brittleness. They report performance, not capacity. As judgment erodes, variability often declines. Dashboards become cleaner. Clean dashboards are read as resilience when they may represent a narrowing perception.

AI cannot remember why leaders once chose to slow down. It cannot recall the tension preceding a failure that never occurred because someone intervened early. It remembers outcomes and forgets vulnerability.

Without governance that actively protects interpretive space, AI becomes a system that maintains performance while erasing the conditions that made resilience possible.

The Constraint That Remains

No system that learns exclusively from its own success remains resilient. Memory without interpretation hardens into confidence. Confidence erases pause. Pause is where judgment lives.

Judgment cannot be preserved by storing more data or building more sophisticated models. It survives only when leadership governs memory itself, deciding what must not be optimized away, encoded prematurely, or delegated entirely to pattern recognition.

That decision cannot be automated.

Succession will reset legitimacy unless leaders deliberately rebuild it. Memory will displace experience unless conditions that create lived interpretation are preserved. AI will accelerate forgetting unless its optimization boundaries are governed explicitly.

These forces do not announce themselves during a crisis. They operate during stability. They work while performance appears strong.

You can design judgment infrastructure. You can protect it through governance. You can scale it across systems. If you do not actively defend it against succession, abstraction, and acceleration, it will erode anyway.

These are not problems to solve once. They are conditions to manage continuously. If you are not managing them deliberately, they are managing your operation by default.

Here's how you know if you're managing them: When was your last succession? Did you explicitly rebuild legitimacy for the new leader, or did you assume it transferred automatically? When was your last near-miss? Did you capture the reasoning that prevented it, or did you just close the ticket? When did you last upgrade your AI? Did anyone ask what interpretive space it was removing, or did you just approve the ROI?

If you can't answer those questions, you're not managing these forces. You're hoping they will manage themselves.

The final question is not whether judgment matters. It is whether you are willing to defend it permanently. Because permanent is the only timeframe that works.

"No strategy is effective without the team. People are the most important thing. You have to make it about them."

— *Laura Phillips, VP of Engineering & Procurement, Pella Corporation*

CHAPTER TWENTY-THREE:

What Never Expires

Judgment does not disappear because leaders stop caring. It disappears when responsibility is treated as time-bound.

Throughout this book, you have seen how judgment erodes under conditions that appear positive. Success narrows variance. Optimization formalizes interpretation into a process. Governance softens as stability increases. Succession resets legitimacy. Memory abstracts experience. AI accelerates confirmation. None of these forces is dramatic. None announces itself as a threat. They operate during performance, not during failure.

What ultimately fails is not awareness. It is ownership sustained over time.

> **"Investing in your current employees is just as valuable as trying to find new ones. If I can take my existing worker and invest in them more - send them for robot training, vision training, automation training - I invest in them to fit that demand."**
>
> — *Jake Hall, The Manufacturing Millennial*

Judgment Cannot Be Delegated

Judgment is frequently described as a human capability, something individuals either possess or lack. In practice, it operates as infrastructure. It determines when a system may proceed and when it must pause. It establishes who holds authority when conditions shift faster than measurable proof can accumulate. It governs the space between confidence and commitment.

Infrastructure does not survive on intention. It does not endure because leaders value the concept. It persists only where authority continues to govern deliberately, especially when visible risk declines and performance remains strong.

If judgment is treated as a capability rather than infrastructure, it becomes discretionary. Once discretionary, judgment is eventually subordinated to flow, efficiency, and scale. Systems are designed to remove friction. Judgment introduces friction by design. Without active governance, the system resolves that tension in favor of momentum.

This outcome does not require negligence. It requires only inattention.

The Illusion of Completion

Organizations frequently behave as though judgment can be installed and then relied upon. Once redesigned at key commitment points, once encoded into governance structures, and once reinforced through training, judgment is assumed to be stable.

That assumption is false.

The pressures working against judgment are continuous. Optimization never stops advancing. Performance incentives never stop narrowing tolerance for delay. Leadership transitions never stop resetting legitimacy. Technological systems never stop abstracting experience into patterns.

Governance, therefore, cannot be episodic. It cannot be a phase of transformation that concludes. It must operate as a standing condition of leadership. The moment active governance relaxes, authority migrates toward whatever is most efficient, most measurable, and least resistant to scale.

This migration does not appear destructive at first. It appears disciplined. It appears modern. It appears data-driven. Only later does the absence of early interruption become visible.

Where Authority Quietly Retreats

Judgment is most visible during disruption. When risk is obvious, leadership intervenes decisively. Authority is clear. Protection is explicit. Interpretation is welcomed.

When stability returns, authority recedes incrementally. Oversight becomes periodic. Intervention becomes exceptional. Meetings confirm performance rather than question direction. The system continues to operate effectively, which reinforces the belief that strong governance is no longer as necessary as it once was.

The retreat does not feel like an abdication. It feels like efficiency.

Yet authority that is not exercised gradually loses standing. Decision rights remain defined on paper, but their practical use narrows. Leaders intervene later in the cycle. Signals that once prompted a pause are allowed to pass. Optimization fills the space authority leaves behind.

Intent does not alter this pattern. Awareness does not neutralize it. Systems respond to design and exercised authority, not to aspiration.

The Burden That Cannot Be Transferred

Judgment endurance cannot be delegated. It cannot be automated. It cannot be embedded once and assumed to be permanent. Every generation of leadership inherits systems that appear stable and must decide whether stability is sufficient.

That decision belongs to leadership because only leadership controls authority design, incentive alignment, and governance cadence across time. No downstream function can sustain judgment infrastructure if executive authority no longer insists upon it.

This responsibility is not symbolic. It is operational. Leaders must be willing to defend early interruption when outcomes do not justify it in hindsight. They must protect interpretive capacity even when optimization produces compelling improvements. They must reestablish legitimacy after succession rather than assume it transfers automatically.

These actions rarely feel strategic. They feel administrative. That is precisely why they are neglected.

What Actually Erodes

Judgment is not displaced by technology alone. It erodes when interpretation is no longer required for the system to proceed. As systems mature, more decisions migrate into defaults. Thresholds tighten. Escalation windows narrow. Variability declines. These shifts are rewarded because they improve throughput and predictability.

Over time, the space in which human interpretation meaningfully alters direction shrinks. Judgment does not vanish abruptly. It becomes unnecessary in routine cases and is therefore exercised less frequently. What is exercised less frequently becomes less practiced. What is less practiced becomes less trusted.

Optimization accelerates this progression by formalizing what has worked. Succession interrupts the relational capital that legitimized early intervention.

Organizational memory retains outcomes but gradually loses proximity to the uncertainty that shaped them. AI reinforces patterns that confirm themselves and filters out what cannot be encoded.

None of these forces requires active rejection of judgment. They operate through continuation.

The Governing Reality

Judgment functions as infrastructure. Infrastructure defines what may proceed and what must pause. Infrastructure that is not actively governed does not collapse dramatically. It thins gradually until the capability it once provided is available only in crisis.

Resilience depends on preserving interpretive capacity before disruption occurs. That preservation requires leaders to govern not only processes but the conditions under which interpretation is legitimate. It requires maintaining authority that can interrupt performance when nothing appears wrong. It requires resisting the impulse to treat clean dashboards as proof of structural strength.

If governance becomes conditional upon visible failure, judgment will always arrive too late.

The Permanent Responsibility

Responsibility for judgment does not conclude when a framework is implemented. It does not expire with tenure, experience, or prior success. It persists as long as the system continues to operate.

The work described in this book is not a transformation. It is governance. It is the ongoing decision to preserve interpretive capacity in an environment that rewards speed, certainty, and automation. It is the discipline to hold boundaries when performance appears to justify relaxing them.

You now understand how judgment erodes, how it can be redesigned, how it must be protected, and how it is lost through succession, abstraction, and acceleration. What remains is not conceptual.

It is structural.

You will be asked repeatedly and routinely whether an interruption is still necessary. Whether authority still needs to pause momentum. Whether friction is still justified. Those decisions will rarely feel historic. They will feel routine.

In those routine moments, judgment either endures or disappears.

What never expires is the responsibility to govern it.

Here's what I need you to understand: This book isn't about AI. It's about leadership that was never taught how to govern what can't be automated.

You were trained to optimize. To measure. To scale. Nobody taught you how to protect judgment when every incentive is pushing you to engineer it out.

So now you're running plants at machine speed, making decisions at machine pace, and hoping, genuinely hoping, that when something completely novel shows up, your people will just know what to do.

They won't.

Not because they're not smart enough. Because you designed systems that trained them not to.

The only question left is this: Are you going to keep treating judgment like it's optional? Or are you going to do the more difficult work and govern judgment as though everything depends on it?

Because it does.

AFTERWORD

Oakridge Industries doesn't exist. There is no Midwestern plant where Denise Tarlow works as Operations VP. No Line 4. No Marta, Devin, or Frank.

But everything that happened at Oakridge is real.

The bearing failure that shut down the line? I heard that story from a plant manager in Michigan. The training room that went dark? Three leaders described the same thing happening in different facilities. The young worker who left after nine weeks because he wasn't learning? I've heard that story dozens of times. The experienced operator who stopped trusting her own judgment because "the system would catch it"? That one came up so often I stopped counting.

Oakridge is a composite: a teaching tool built from documented patterns across dozens of real organizations. The characters are fictional. The dynamics are not. The facility is invented. The crisis is epidemic.

If you saw yourself in Oakridge's story, it's because the pattern is widespread, not because I'm describing your facility specifically.

Where This Comes From

Between 2020 and 2025, I conducted more than 200 long-form interviews with manufacturing leaders through the Manufacturers' Network podcast. These weren't sound bites. They were twenty- to forty-minute deep-dives into the challenges facing modern manufacturing.

The guests included CEOs and plant managers of small to mid-sized manufacturers, operations directors and supply chain leaders, HR and talent development professionals, former executives from Amazon, major automotive suppliers, and global manufacturers, manufacturing consultants and technology providers, industry association leaders, and next-generation workforce advocates.

When transcribed, these conversations totaled nearly one million words – roughly the length of ten typical business books.

Every quote attributed to a guest on the Manufacturers' Network podcast is used with their knowledge and permission. Where guests requested confidentiality about specific company details, I've honored those requests while preserving the insights. The patterns you've read about in this book emerged organically from those conversations, often without prompting.

What's Coming (And Why Most Plants Aren't Ready)

You deserve clarity, not clichés. So, here are the truths most manufacturers aren't ready to say out loud:

Prediction #1: The leadership crisis will hit before the labor crisis ends.

Robots can weld, pick, place, and inspect. They cannot lead. And most plants have no viable next generation of leaders in development.

Prediction #2: Small and midsize manufacturers will become talent farms for big brands.

If you do not intentionally retain your people, you will train them for your competitors.

Prediction #3: The first wave of AI "success stories" will collapse from skills starvation.

Plants that automate the fastest without reskilling will suffer the longest, most costly outages.

Prediction #4: Your future depends on your ability to reskill mid-career workers, not recruit new ones.

The workforce you need does not exist yet. You must build it.

Prediction #5: Owners who treat AI as a technology project will fail.

Owners who treat AI as a workforce transformation will win.

The Fork in the Road

Every manufacturer, large or small, is standing at the same fork:

Path One: The Hollow Factory

This is the path of least resistance. Deploy technology. Cut costs. Assume the workforce will adapt. It leads to a hollowed-out culture where nobody feels ownership, evaporating skills as knowledge walks out the door, and a leadership vacuum as career paths disappear. Turnover accelerates because the best people leave first. Technology gets stranded when nobody knows how it works.

Catastrophic failures happen when perfect systems meet imperfect reality. You end up with a facility that's technically advanced but structurally unstable. Most manufacturers think they're avoiding this path, but here's the thing: If you're not actively building an alternative, this is where you'll end up by default.

Path Two: The Smart Plant™

This is the intentional path. It requires vision, investment, courage, and a willingness to challenge conventional wisdom about what automation actually means. It leads to technology that amplifies human capability rather than replacing it, a systematically reskilled workforce that grows more valuable over time, and a high-trust culture where people and machines work in genuine partnership. You get sustainable automation that makes work better, not just cheaper.

This is the plant that survives for the long run. The one that customers trust. The one where talented people want to work. The one that makes money consistently because it can handle whatever reality throws at it. You get leadership pipelines that produce the next generation of plant managers and competitive advantages that compound year after year, a facility that's both technologically advanced and fundamentally resilient.

But you only reach this destination through deliberate design.

The manufacturers who figure this out first won't just survive the next twenty years. They'll dominate their markets.

ACKNOWLEDGMENTS

This book exists because more than 200 manufacturing leaders were willing to talk honestly about what keeps them up at night. To every guest who joined me on the Manufacturers' Network podcast between 2020 and 2025, thank you for your candor, your expertise, and your willingness to share not just what works, but what worries you. Your voices form the foundation of everything written here.

Special gratitude to those whose insights appear as quoted voices throughout these pages. Each quote represents a moment when someone crystallized a complex problem into a truth so clear that it stops a reader mid-page.

To Randy Gage and Roger Atkin, whose words appear on the cover of this book – your insights cut to the heart of what this book is really about: what human thinking is worth, and why the pace of change makes protecting it urgent.

To Alex Ladd and Karla Trotman, you made visible the gap between what dashboards measure and what matters.

To Eknauth Persaud, Lisa Sanderson, Jason Vanzin, and Lee Rector, you clarified what AI can and cannot do, and what gets lost when automation is mistaken for capability.

To John Kevin Koehler, Katie Smith, Brent Kedzierski, Noah Graff, Darrin Mitchell, Vinny Maurici, and Dave MacDonald, you showed what judgment infrastructure looks like when it is protected, and what erodes when it is not.

To Tom Hatton, Joe Ricci, Jim Ver Woert, and John Ballinger, your stories reinforced that manufacturing is fundamentally human work, and that no product, however excellent, can substitute for a strong culture.

To Mike Bowers, you brought clarity to the technical complexity that makes this challenge harder than it looks.

To Jake Hall, Will Healy III, Lori Anderson, Ivan Madera, and John Wilczynski, you represent the forward momentum of this industry. Your perspectives on technology, talent, and leadership shaped the future-facing edge of this book.

You didn't just contribute to this work. You made it real.

To Vince Sassano, whose observation – that leaders were treating digital transformation like a tech install instead of a governance decision – sparked the chain reaction that led to this book, you saw it first.

To Roger Atkin and the NTMA, thank you for trusting me with conversations that deepened my understanding of precision manufacturing and machine tooling.

To the operators, technicians, supervisors, and plant managers I've met in mills, shops, and factories across the country, you are the invisible infrastructure of modern life. Your judgment, skill, and attention make the world work. This book is an attempt to make visible what you've always known.

My gratitude to the Mastermind communities that sharpened this work from a pattern into a framework.

To my Cigar PEG Mastermind group: Bryce Austin, Carolyn Strauss, Ed Rigsbee, Erik Larson, Kelly Vrla, Ken Blackwell, Mike Wright, Mitch Seehusen, Scott Cooksey, Tom Guetzke, and Vivian Cobb. Thank you for helping me find the spine of this book when it was still a loose collection of thoughts.

To my National Speakers Association CSP Summit Mastermind Group: Carolyn Strauss, Florian Ilgen, Frank Kitchen, Katherine Belt, Molly Wendell, Sally Spencer-Thomas, and Scott Friedman. Your insight and challenge pressure-tested the thinking and strengthened the argument.

To my KK5 Mastermind family: Debbie Peterson, Justin Patton, and Ted Ma. We've been meeting since 2018. You are not just colleagues; you are constant anchors. Thank you for your clarity, support, and insistence that this book live up to its responsibility.

To Dale Evraets, my boss during my years in the welding industry, thank you for teaching me to listen to the people on the floor first. That lesson shaped everything that followed.

To the readers of Gratitude Thought of the Week and the Cracking the Retention Code community, thank you for staying in the conversation as my thinking evolved.

To Vicki McCown, thank you for your sharp line editing and the care you brought to every sentence. This book is clearer, more precise, and more readable because of you.

To Shaina Nielson, thank you for your formatting precision and for shepherding this book through production.

To my clients over the years, for every challenge you shared and every system you questioned that informed this book. You've taught me as much as I've ever taught you.

To my mom, Roslynd Smith, thank you for your constant support.

To my husband, Scott Ryan, thank you for the space, patience, and belief required to see this through.

ABOUT THE AUTHOR

Lisa Ryan, CSP, MBA, is Founder and Chief Appreciation Strategist of Grategy and host of the *Manufacturers' Network* podcast. Since 2020, she has conducted more than 200 long-form interviews with manufacturing leaders, operators, and technology providers. Those conversations, totaling nearly one million words of transcribed insight, form the foundation of *Smart Plant*™.

With more than thirteen years of experience in industrial sales, including seven in the welding industry, Lisa has worked in steel mills, fabrication shops, automotive plants, and food processing facilities across the United States. She understands manufacturing not from conference stages or consulting frameworks, but from the plant floor, where decisions are made under pressure and judgment carries real consequences.

Lisa is a keynote speaker, consultant, and advisor to manufacturing, construction, and skilled trades organizations, helping them navigate workforce retention, leadership accountability, and the human implications of automation. Her work focuses on designing systems that allow people to think, interrupt, and exercise judgment, not just follow dashboards.

She is the author of twelve books on leadership, retention, and workplace systems, including *The Upside of Down Times: Discovering the Power of Gratitude*, featured in Harvey Mackay's syndicated column. Her book, *Manufacturing Engagement: 98 Proven Strategies to Attract and Retain Your Industry's Top Talent*, is widely used by manufacturing leaders to address workforce instability.

A Certified Speaking Professional (CSP), a designation held by fewer than 17% of members of the National Speakers Association worldwide, Lisa is also a Past President of the National Speakers Association Ohio Chapter. She holds an MBA from Cleveland State University and has been recognized as Corporate Event Speaker of the Year and inducted into the Virtual Speaker Hall of Fame.

While her earlier work focused on gratitude and retention, *Smart Plant* reflects a deeper shift in her thinking toward governance, authority, and human judgment in automated systems.

Lisa lives in Cleveland, Ohio, with her husband, Scott Ryan. She continues to interview manufacturing leaders and advise organizations on how to design systems that preserve human judgment as automation accelerates.

CONNECT WITH LISA

Manufacturers' Network Podcast: Manufacturers-Network.com

Website: LisaRyanSpeaks.com

Email: Lisa@Grategy.com

LinkedIn: LinkedIn.com/in/AskLisaRyan

A NOTE BEFORE THE ASSESSMENT

This book makes a claim: that judgment is not a soft skill. It is infrastructure. And like any infrastructure, it either exists by design or erodes by default.

The assessment that follows is not a test, a scorecard, or a measure of leadership intent. It is a diagnostic lens. Its purpose is to make visible where judgment is currently protected, permitted, and governed inside your organization, and where it is being overridden by speed, metrics, or certainty.

Answer the questions based on how your systems actually operate, not how you wish they did. High scores do not mean you are finished. Low scores do not mean you have failed. They simply reveal where judgment is being designed in and where it is being designed out.

That clarity is your responsibility.

Access the Smart Plant Framework™ Self-Assessment

https://lisaryanspeaks.com/smart-plant-frameworkself-assessment/

www.ingramcontent.com/pod-product-compliance
Ingram Content Group UK Ltd.
Pitfield, Milton Keynes, MK11 3LW, UK
UKHW021523300726
14060UKWH00018B/753/J

9 798995 174707